Don't Let Your GEOLOGY Mess Up Your THEOLOGY

Douglas B. Sharp

EABooks Publishing
Your Partner In Publishing

Quotations from the Holy Bible are from the King James Version unless otherwise indicated.

Revolution Against Evolution
2054 Valley Forge St.
Grand Rapids MI 49504
(517) 321-2129
https://www.rae.org

Because of the dynamic nature of the Internet, any web addresses or links contained in this book may have changed since publication and may no longer be valid. Photographs are by the author, Guy Forsythe, David Vonderheide and Linda Rusiecki.

ISBN: 978-1-963611-16-8
LCCN: 2024908206

Published by EA Books Publishing, a division of
Living Parables of Central Florida, Inc. a 501c3
EABooksPublishing.com

TABLE OF CONTENTS

Endorsements

Don't Let Your Geology Mess Up Your Theology by Doug Sharp is a *tour de force* of some of the most stunning geology in America. He takes the reader on a sightseeing walk explaining the implications of the beauty of the landscape revealed by the walk. Doug's stunning photographs help us to fully appreciate God's creation.

The study of rocks does not sound very romantic, as does the study of emeralds, sapphire, amethyst, topaz, ruby, and diamonds. These, and many other precious stones, are nothing but polished rocks, as are granite, marble, and slate. In the rough, most rocks look worthless. But when polished they become things of exquisite beauty. Sharp, in his tour, helps to open up the door to their beauty. The Bible mentioned minerals in 43 scriptures. One example is the twelve sacred gemstones of Revelations, which are jasper, sapphire, chalcedony, emerald, sardonyx, sardius, chrysolite, beryl, topaz, chrysoprase, jacinth, and amethyst. Many Bible scholars also believe that there is a clear association between the twelve apostles and the twelve gems from the walls of Jerusalem.

Your tour guide is a creationist who will point out the evidence of creation instead of evolution in the world pictured for us in his walking tour of the West. The scriptures declare the heavens reveal the glory of God, and this includes the sky, the mountains, and the hills. Anyone who loves nature will love this informative, educational, and entertaining book.

Jerry Bergman, Ph.D.

Acknowledgements

For the foundation of this book, I credit my friendship with creationist geologists Guy Forsythe of Crying Rocks Ministry in Sedona and David Vonderheide who hosted us in Flagstaff several times while we explored the geology of Arizona. Also my Canadian friend Ian Juby with whom I explored the Joggins Cliffs and the Niagara Falls area. I appreciated David Oberpriller, who conducted Plants of the Bible tours at Boyce Thompson Arboretum in Superior, Arizona and took us to the Chinese Wall at South Mountain.

Thanks goes to Dr. Joseph Kezele of the Arizona Origin Science Association. Michael Oard and Kevin Anderson gave us a tour of Glacier National Park. Dr. John Baumgardner, Dr. D. Russell Humphreys, Dr. Stephen Austin all gave us interviews, and Allen Roy gave us another viewpoint at the Grand Canyon. We were also able to interview Dr. Henry M. Morris and his son John Morris, whose pioneering work with the *Genesis Flood* book and establishment of the Institute for Creation Research started the revival of creationist research.

I honor Ken Ham for his vision to communicate the message of God's creation at the Creation Museum and Ark Encounter. His crew of scientists include Dr. Danny Faulkner, Dr. Terry Mortensen, Dr. Andrew Snelling, and Buddy Davis. The ministries of the Creation Research Society, Creation Moments and Creation Ministries International all are excellent sources of information.

Special thanks goes to Dr. Jerry Bergman, who co-authored two books with me, *Persuaded by the Evidence* and *Transformed by the Evidence*. We had many creation scientists supply us with their personal testimonies for those two books.

Dr. John Sanford, author of *Genetic Entropy*, invited me to participate in the GENE conference in San Diego. Dr. John N. Moore was my professor at Michigan State University who had the courage to team teach natural science with an evolutionist and present the creation side. Dr. Carl Baugh gave us a personal tour at his Creation Evidence Museum and showed us all of his out of place artifacts and dinosaur tracks. Don Patton supplied me with a lot of useful information in his PowerPoint presentation. The late Dr. Raymond Damadian, inventor of the MRI, graciously granted us interviews with him at his company.

I honor my loyal friends who supported me for many years as my TV crew for the Revolution Against Evolution show and accompanied me on our adventures. They are my co-host Rich Geer, Larry Perry, David Page, Ted Johnson, Eric Armbrustmacher, Jon Iversen, Freddy Torres, Derek Marshall, Sue Langham, Dan Janzen, and Jeff Conklin. Revolution Against Evolution is perhaps the longest continuously running creationist TV show, on the air since 1993.

Finally I credit my longsuffering wife Vivian, who wasn't that interested in looking at rocks but supported me faithfully as I pursued my geologic adventures. She is now with Jesus, and I have a book about my life with her called *No Worries in Heaven*.

Introduction

The intent of this book is to explain the Biblical perspective on the origin of the earth and its geological features. The Bible clearly has a lot to say about this subject, and for centuries the stories in the Bible were treated as fact and revered by eminent scientists like Nicholas Steno and Isaac Newton as the foundation for scientific truth. During the eighteenth and nineteenth centuries, people were seeking to throw off the oppression of monarchies who justified their right to rule from the teachings of the church and the Bible. To do this, they sought to discredit the authority of the Bible as the word of God by introducing the concept of evolution over long periods of time.

The divine right of kings was a concept certainly worthy to be debunked, but rather than addressing the issue by revealing that the Bible didn't really teach that, Scottish geologist James Hutton and later Charles Lyell undermined the authority of the Bible by proposing that the layers of rock strata represented millions of years of accumulation. Later, Charles Darwin built upon this by proposing the theory of evolution.

The church was not prepared for this and responded with compromise theologies like the day-age theory and the gap theory rather than challenge the assumptions and motivation behind the evolutionary theory.

Don't Let Your Geology Mess Up Your Theology is a book that explains how you can read scripture in a straightforward manner and at the same time use it to explain the features you see in the geologic record. Much has been written about this from a creationist perspective, but some of those books can be ponderous, technical, and difficult to digest. This book starts with scripture, shows you where clear choices must be made concerning what you believe about God, and why it is important. Then the goal is to teach you what to look for as you observe geological features. Evolutionists and creationists have the same scientific evidence, but their starting assumptions drive their conclusions.

Most of the later chapters are examples of where you might travel to view geological features and understand how what you see fits within a creation framework based on the Bible. I visited most of the places mentioned in this book, and I can tell you how to get there and what to look for.

As you read this book, I encourage you to consider the God of the Bible and ask yourself if He is true or false. If He is true, is He personal, does He care about you, and if so, would He lead you astray? If He is the creator of the world and created you, He is perfectly capable of revealing to you the truth and able to overpower any other lying voices that may be out there. Connect with Him and tell Him that you want to know the truth. He will answer that prayer.

CHAPTER 1

The Geology Story in the Bible

One of the most definitive statements about the origin of the earth is in the Holy Bible, and it is the very first verse:

In the beginning, God created the heaven and the earth. (Genesis 1:1)

If we believe that this verse is true, it flies in the face of much popular philosophy and what is taught in universities and schools. It refutes evolution, materialism, historical geology, and atheism. It refutes the idea of the origin of life from chemistry. It refutes uniformitarianism, which is the story that the "present is the key to the past", and that what we see in the geologic record took place over millions of years.

This verse also places God at the center with a direct hand in creating earth with life on it. Some philosophies, like Deism, place God in the distant unknown past, talk about Him starting things up, but leaving it alone to evolve on its own. That's not what the first verse of the Bible implies.

If you read the next few verses in the first chapter of Genesis, you will understand that it is describing the process how God created our world, and it took place very quickly, over a period of six days, with the seventh being the day of rest. There have been many attempts by scientists, skeptics, and theologians to integrate long ages into the Genesis story account, but none of these work. In the Ten Commandments, there is a verse in the Bible describing the creation week:

For in six days the LORD made heaven and earth, the sea, and all that in them is, and rested the seventh day: wherefore the LORD blessed the sabbath day and hallowed it. (Exodus 20:11)

This verse corroborates the full meaning of the first verse in the Bible and establishes the fact that the Bible describes the creation event occurring over the course of a week. This excludes the idea of a long gradual course of evolution. The evolutionary explanation is impractical. The creation needed to take place quickly in order for the balance of integration of all forms of life to take place. The traditional seven-day week is established throughout the history of mankind and doesn't make sense except in the light of the Biblical origin account.

Some theologians attempt to place millions of years of evolution between the first verse and second verse, saying that there was a period of time where the earth was ruled by Satan and there was a great cataclysm where he was kicked out of heaven, and this is where God is recreating the earth. This is the "gap theory". The problem is that it is an argument from silence. The Bible says nothing about that and if that were the case, He would have included that important information right here. He didn't. Verses 1 and 2 are connected with an "and" and that strongly implies that they occurred in sequence with no gap in between.

On the first day, God created light and divided the light from the darkness. He called the light Day and the darkness Night. Then comes a powerful statement:

And the evening and the morning were the first day. (Genesis 1:5b)

This statement is repeated in verses 8, 13, 19, 23, and 31 for each of the days of creation. This refutes the "day-age theory" that claims that the days of creation were periods of evolutionary development over millions of years.

The second day talks about the creation of heaven (firmament) and the dividing of the waters above from the oceans below.

The third day describes the creation of the dry land, grass, herb, and fruit trees. The majority of the geologic formation of the basement rock occurred in these two days. God then first created plants as food. This was necessary for life to thrive in the newly created world. It was in the right sequence.

On the fourth day, the sun, moon, and stars were created. Before this occurred, the light originally created by God illuminated the earth. Now the light was appointed to be the sun and the moon ruling the day and night. This perhaps may be the most discussed verse amongst Bible believing creation scientists, as it involves the starlight and time problem. Discussion of this subject is beyond the scope of this book, but creationists have many different intriguing proposals on how to explain this.

On the fifth day, God created sea life, fowl, great whales, and every living creature that moves. Then on the sixth day, God created the rest of the living creatures including man.

The story of the creation account in the Bible certainly involves a lot of miracles. A miracle is by definition something that is outside of our understanding that does not follow the normal rules of the universe. It is non-testable, not repeatable, and by definition something we are unable to use the tools of science to verify.

However, as you will discover in this book, to dismiss the creation account in the Bible, is to believe in something else that is far more miraculous and difficult to believe, such as the chemical origin of life from non-life. The theories that are proposed by non-believers and atheists are not subject to scientific test and are not repeatable. In fact, chemical origin of life experiments fail. Since we cannot go back in time to discover firsthand what happened, we need to rely upon written historic accounts from someone who was there and had a hand in it. That is God who inspired the Biblical account. It is our choice which set of miracles we want to embrace.

The account of creation in the beginning has a small part in understanding the geology of the earth. It explains the bedrock and the nature of our planet with all of its life forms that existed before the great flood of Noah.

Most popular and conventional textbooks on geology exclude the possibility of the worldwide flood described in chapter 6 of Genesis. The doctrine is that the present is the key to the past, and that by examining present processes we can understand what happened over the course of millions of years of earth history. Uniformitarianism is what this doctrine is called, and it is an assumption, not something that is established by scientific observation.

A good illustration of this I like to use is the Lansing Mall in Lansing, Michigan. If you send a team of scientists to explore the grounds and dig up evidence that will describe what it was like 60 years ago, and from that evidence paint a picture of what if might have looked like, I am sure that the picture the scientists would come up with would be quite inaccurate. Since that property was my grandfather's farm, and I know its history and grew up there, I can firsthand describe where the farmhouse and barn stood, the garage and milkhouse, the garage, chicken coop and water trough. By digging in the rocks, how much tougher would it be for the evidence scientists retrieve to describe the geology of landforms and conclude that they came into existence over millions of years? Empirical science is where you can conduct an experiment, repeat it, and establish it by observation. Forensic science takes what you find in the present and uses that to guess what occurred in the past. It does not have the same weight.

What do we know about the great flood of Noah? First, it was worldwide and covered the mountains.

And the waters prevailed exceedingly upon the earth; and all the high hills, that were under the whole heaven, were covered. Fifteen cubits upward did the waters prevail; and the mountains were covered. (Genesis 7:19-20)

This was enough to destroy all life (verses 21-23). If it were only a local flood, it would not have done so. Mountain building took place during the last stages of the flood, so initially the waters covered the pre-flood mountains. Mount Everest has marine fossils on top, it was post-flood.

The waters prevailed upon the earth an hundred and fifty days. (Genesis 7:24).

It is the consensus of those who believe in the Biblical account of creation that this worldwide flood explains what we see in the geological record. Up to the so-called "enlightenment" in the 1700s, that was the perspective that was taken by most men of science before that like Isaac Newton and Nicholas Steno who honored God with their scientific investigation. We will see how philosophy and politics of that time influenced the thinking of scientists to abandon the Biblical timeframe and the morality of the Bible as well.

CHAPTER 2

Creation Geology Explained

The Bible is the best-selling book in the world. It makes the claim that it is a record of the history of the world from its beginnings. It is the foundation of Christianity and Judaism, and it has been revered for centuries as an authority in matters of life. The principles taught in the Bible work. It has survived tests from archaeology and science.

We have a choice. We can believe that it is true, or we can believe that it is false. If we believe that it is true, we can believe the straightforward narrative found in the Bible, or we can think that the wording in the Bible means something other than the direct meaning. Some may also argue that parts of it are true, but some of it is myth. They do so to make peace with those who believe that it is false, hoping to find intellectual solace in this compromise.

Those who believe the Bible is false have crafted a multitude of alternative theories about the origin of the earth and life on it. Most of these theories fall under the general category of "evolution" and seek to explain the origin of the world from a naturalistic and materialistic viewpoint. I would argue that these ideas are ideology driven, arguing from the assumption that there is no God, and the Bible is false. It is not surprising that if you start out with the premise that there is no God or that He is so distant in the past that he isn't worth considering, that anything that follows logically will end up with the same result. We will deal with this more in chapter 3.

We believe that the Bible is true, and that it ought to be understood in a straightforward way. This affects how we view geology. There are several Biblical events that shaped our world, and these are understood clearly from the text. These are:

- The creation of the earth and its landforms on the second and third days of the creation week.
- The period between the Garden of Eden and the start of Noah's Flood where life abounded.
- The fountains of the deep were broken up and the windows of heaven were opened (Genesis 7:10-11).
- It rained for 40 days and 40 nights.
- The waters prevailed on the earth and covered the high hills and mountains. (Genesis 7:18-20).
- In this flood, all of the creatures on the earth except those in the Ark died and were buried in the flood deep in the sediment forming layers of coal, oil, and gas (Genesis 7:21-23). These layers average about 900 feet deep and can be up to 12,000 feet deep.
- The flood lasted 150 days (Genesis 7:24).
- The waters returned off the earth after 150 days (Genesis 8:1-5).
- There was a period where the earth was drying out (Genesis 8:13-14).
- The aftermath of the flood resulted in a new weather pattern that we believe created massive glaciers and ice in the northern and southern parts of the world.

For those who believe the Bible is true, we have a choice. We can believe that the account of the flood in Genesis is true, or we can believe that it is false.

For those who believe it is true, we can believe that it was worldwide and covered the entire earth or believe that this was describing a local flood.

The local flood theory has its problems. If God was just going to flood part of the world, why did He instruct Noah build an ark? He could have simply told Noah and his family to move to higher ground. God made a promise never again to flood the earth (Genesis 9) and established the rainbow as His sign. Certainly there have been local floods of great size since, but not on the global scale as is clearly described in Genesis.

If you consider the creation of the world and know that the global flood was worldwide, it provides with it a powerful amount of scientific explaining power for what we can see in geology. We will now explore how that is the case.

The formation of the earth during creation week describes what the world was like before the flood. The waters were gathered into one place (Genesis 1:10) and the dry land appeared. There was one continent. The mountains and hills would have been small. Most of the dry land would have been made up of primordial igneous rock like granite and basalt. The nature of granite is such that it has never been reproduced in the laboratory, so we would call it our creation rocks. When scientists try to synthesize granite they get rhyolite. The importance of radioisotopes in granite to the creation model will be discussed later.

The original created world would have been a lush tropical environment built with the purpose of lasting for eternity. The firmament described in Genesis 1:7 would have provided a measure of protection from cosmic rays and yielded long life spans for man and the creatures in the world. It is not surprising that we find giant plants and animals as fossils from that world that we do not see today. Adam lived to be 930 years old, Noah 950 and Methuselah 969 years old. We would expect that the world's atmospheric conditions and environment would be much different from what we observe today. This was also the time of the dinosaurs. There would

have been also a much greater genetic diversity. Not only would there have been a much wider variety of animals, but the genome would have been fresh from the hand of the creator and much more robust.

It is well known that the genome is deteriorating and that we are losing species. New species of animals are not appearing, but extinction is occurring at a rapid rate. This is consistent with the idea that our creator built into the genetics of the system of life mechanisms to preserve it, but we also know because of the sin and death in the world, that the genome doesn't contain all of what it had in the beginning.

After the creation of the world, Adam and Eve enjoyed all kinds of freedom in the Garden of Eden. They were given a test and were told that they were not to eat of the tree of the knowledge of good and evil. They chose poorly and failed the test. This is the doctrine of original sin, and it is the reason Jesus Christ came to redeem us.

With the account of original sin we have a choice. We can believe that this is historically true, or it is false. Some believe that mankind is inherently good, but it's his environment that causes him to be bad. This diverts the blame for man's sin to God and His creation and brings the excuse that it's God's fault for all of the evil in the world.

The opposite is true. Man is responsible for sin, and the environment suffers as a result. It also brings us to another belief choice. Is the devil real, or not? The Bible clearly talks about him as the tempter in the Garden of Eden and he is the source of all of the evil in the world. If the devil is real, we have another choice. Do we resist him or not? How we respond to the temptations in life corresponds to how much trouble we will receive. Experience with this is sound proof that the Bible is the word of God.

There came a point where the wickedness of man caused God to consider destroying what he made to start all over with the one man he found to be righteous. Noah found favor in the sight of the Lord and God instructed him to build an ark. Noah's ark was a massive structure, and it was built to house all of the animals representing all of the genome types safely during the coming cataclysm. A representation of Noah's Ark to scale is built as the Ark Encounter in Williamstown, Kentucky.

We can choose to believe the Biblical account of Noah's Ark to be true or false. Skeptics believe that the Ark wouldn't have been adequate to house all of the millions of species known on the earth today. But that confuses the meaning of our arbitrary definition of species with the Biblical kinds. God places the creatures on the ark according to their ability to breed, and the genetics carried by those representatives on the ark branched out into the varieties we see today on the earth. This could be argued to be a kind of "evolution" but that is merely a diversification of species based upon the original kinds. Random mutations do not result in new information, but God designed within the genome tremendous variety.

Creationists roughly equivocate the Linnean family classification to the Biblical kind. The construction of the Ark, its seaworthiness, how Noah fed the animals, got rid of the waste, whether dinosaurs were on the ark are all questions skeptics bring up and have been answered in detail. The Ark Encounter theme park in Kentucky is a good place to visit to answer those questions.

The most popular model for what happened geologically during Noah's flood is described as Catastrophic Plate Tectonics (CPT). The Bible says that the fountains of the deep were broken up and the windows of heaven were opened. Several other models have been proposed, and all of them operate from the premise that the source for the majority of the water of the flood was subterranean, accompanied by the collapse of the waters above. Dr. John Baumgardner was one of the first to propose this model, and he based it upon a computer model he built called TERRA. Originally the dry land was gathered all in one place. The CPT model splits this original continent into pieces at the spot where the Mid Atlantic Ridge is, releasing fountains of water supersonically into the atmosphere and that water covered the continents.

Then the continents began to move away from the ridge as plates, and where they collided with the other plates, runaway subduction occurred, drawing the continental mass into the mantle. Evolutionists have their own version of this where the process is gradual, but this CPT subduction model provides a driving force for rapid separation of the continents. The estimated speed of continental movement was 45 miles per hour as the mantle drew the plate downward.

Continental movement of this scale would have resulted in sedimentary rock miles thick, massive uplift, and mountain building. There would have been billions of fossils buried in rock strata all over the earth. In North America, it would have been the Appalachian Mountains, then the Rocky Mountains. This is what we find.

The ocean basin near the Mid-Atlantic ridge is volcanic basalt. Sedimentary rock is not that common beyond the continental shelves and what is found in major river deltas like the Mississippi is meager. In comparison to the thick sedimentary strata on the continents, it is the opposite of what you will predict to find if sedimentation takes place over long ages.

Fossil burial would have been in an order in relation to its proximity to the water. You would find marine fossils throughout the geologic record, fish next, then amphibians, reptiles, birds, and mammals. 95% of fossils are marine invertebrates, then 4.5% are plants and algae. 0.125% are vertebrates, mostly fish. It is a mistake to assume that the representation of the traditional geological column in textbooks accurately represents what is found in the rock strata. We will be giving examples where that is not the case: missing strata, out of order strata and fossils found in the wrong strata.

Mountain building took place rapidly. What was once the sea floor would be pushed up to become the continents and as the Mid-Atlantic Ridge erupted and spread, it would collapse to become the sea floor. Eventually the catastrophic movement of plates slowed, then stopped. We still have residues of this catastrophe occurring near the edges of the plates as earthquakes and volcanoes. It is interesting that thousands of feet of sedimentary rock is piled up on the continents, but there is relatively little of that on the sea floor in comparison.

Rapid burial of vegetation and floating log mats depositing layers of bark and buried under sediment would have produced coal seams.

The new oceans were very warm from all of the volcanic activity and created an environment where the weather was temperate at the poles, but in the interior generated a snow engine that piled up vast glaciers that lasted for hundreds of years after the flood. The temperate climate in the Arctic allowed for life such as mammoths, mastodons, and dinosaurs to thrive, only to be buried, quickly frozen and fossilized in postflood cataclysms. Before the flood, the Bible records the ages of the patriarchs to reach over 900 years, and we can also speculate that the rest of the species of the earth also enjoyed the same kind of longevity. This allowed them to reach great size.

There are numerous geological features that can be attributed to post-flood geologic activity. Though the movement of the plates slowed significantly, there would have been mountain building, volcanos erupting, ice buildup and melting, and canyons forming from the flood runoff.

After the flood, conditions for long life changed, and there could be various reasons for this. The ages of post-flood humans dropped rapidly in the recorded history of the Bible. Many of the species that survived the flood aboard the ark, such as dinosaurs, no longer had the environment that fostered long life. Two possible reasons are the genetic bottleneck at the flood, and that there may have been a protective atmospheric layer that no longer existed.

A lot of water would have been contained in continental ice packs and glaciers, lowering the sea level, and creating land bridges between the continents.

God caused confusion of the tongues at the tower of Babel and as a result, dispersion took place. The land bridges between Asia and North America and Asia through Indonesia and Australia would have existed at the time when the waters were warm, and the coastline had temperate climates. This would have made the population of the new world possible. In addition, the patriarchs after the flood would have carried their ark-building knowledge forward to their offspring, and in the early years their voyages to the new lands would have added to the repopulation of the earth as well as distributing species of plants and animals.

Who hath measured the waters in the hollow of his hand, and meted out heaven with the span and comprehended the dust of the earth in a measure, and weighed the mountains in scales, and the hills in a balance? (Isaiah 40:12)

It is undeniable that the earth is finely tuned to support life, and it is the earth only in our knowledge of our solar system and other planets that exhibits those characteristics. The creation displays myriads of intricate interrelationships based upon key principles of science. The oceans, lakes, rivers, ground water and atmosphere are in precise balance in a system called the hydrological cycle. The rotation of the earth on its axis at an angle of 23 ½° is exactly what is needed to warm the earth at a temperature to support life.

We presume that the environment before the flood was engineered even more optimally is such a way that man had lifetimes almost a thousand years. The earth, sun, and moon system is so precise that the sun and moon appear the same size in the day and night sky and is the only place known where a total eclipse can occur. The earth covers the moon completely during a lunar eclipse and the moon covers the sun completely during a solar eclipse. This is remarkable.

CHAPTER 3
Geology Exhibits Amazing Beauty

If you stand on the rim of the Grand Canyon watching the sunset or gaze through the mist over the Na Pali coast on the island of Kauai, you would be quite a cynic to refuse to give glory to God. Even if you realize that these geological feature are remnants of the great flood, the fact is that there is a tremendous measure of beauty in the artistry of colors. Beauty is not a quality that is measurable scientifically. It is something that is self-evident to the eye of the beholder. I believe that if something is truly beautiful, it is because it brings us into remembrance of the Garden of Eden.

Beauty is outside of ourselves, much greater than our capacity to comprehend. Beauty tugs at our soul and reminds us of the ultimate purpose for which God created us. If something is beautiful, it brings us peace, tranquility, and a sense that this is the way life is supposed to be.

Beauty imitates our creator and if we create a piece of art that is beautiful, it is because it reflects the glory of God. Unfortunately much of what passes

Grand Canyon Desert View

for postmodern "art" reflects the ugliness of the world around us. Those artists believe that we are just a product of chance, random processes and if you don't like their art, well, tough. "We're just reflecting what we believe is reality." I believe this to be a mistake, and if they want to imitate the true reality, an artist needs to consider the wonders that are found in this world that are beyond clinical inspection and beyond our meager efforts to explain.

Pu'u O Kila Lookout, Na Pali Coast, Kauai, Hawaii

Beauty is hard to explain, much like love, faith, truth, peace, and joy. These cannot be analyzed in a laboratory test. Even in the geology of the world that has been ravaged by a flood, the remains of that world shows beauty, reminding us that God in His provision is able to show His glory and care for us in the aftermath of his wrath and destruction when He judged the world. He provided the rainbow as a sign that He would never again judge the world in that manner.

If we approach our study of this world from a Biblical worldview, we are able to gain a perspective that transcends scientific reason. The materialistic worldview that is taught in our schools is missing a spiritual dimension and when that is left out on purpose, it leaves a sense of emptiness.

Double Arch, Arches National Park, Moab, Utah

Seney National Wildlife Refuge, Upper Peninsula, Michigan

On the path to Queen's Bath below Hala Falls, Kauai, Hawaii

Cabin in Carcross, Yukon Territory

CHAPTER 4

Origins of Evolutionary Geology

Great scholars of the past like Nicolas Steno and Isaac Newton revered the Bible as the word of God. Steno (1636-1686) developed fundamental principles of geology that are continuously used to interpret sedimentary rock layers. He noticed that marine fossils like shark's teeth were found far inland, and he realized that they were remains of ancient sharks preserved in rocks once laid down by the sea. He said, "We learn from Holy Scripture that all things, both when Creation began and at the time of the Flood, have been covered by waters."[1] Steno observed that each bed of sedimentary rock was deposited on a solid substratum, and they occurred on horizontal planes with younger strata on top of older strata. Folding and breakage occurred later. This method of stratigraphic interpretation is still in use today and is the foundation of geology.

By tracing the chronology of the patriarchs in the Matthew and Luke accounts, they were able to estimate the date of creation to be about 4000 years before the birth of Christ. Newton's book on Biblical chronology took him forty years and it was only published after his death in 1728.

Archbishop James Ussher (1581-1656) wrote a complete history of the world in Latin, covering every major event in human history from the time of creation to AD 70. He published this 1,600 page volume in 1650, and the English translation was published in 1658. He pinpointed the date of creation to be October 23, 4004 BC. He used the chronologies of the Hebrew text of Genesis 5 and 11 as well as many other Bible passages to arrive at his conclusion. This timeline is tight, and if you take it in a straightforward manner, the margin of error is fairly small. Some modern chronologists may disagree with

Ussher's conclusions by a small percentage. Ussher uses the Masoretic text to establish this timeline. The list of kings given as the genealogy of Jesus traces his descent back to Adam, and the Old Testament books of Judges, 1 & 2 Samuel, 1 & 2 Chronicles, 1 & 2 Kings give enough details to establish a continuous timeline. The times of the Judges and the Persian empire were the areas that exhibited the most uncertainty. This establishes the Biblical age of the earth at approximately 6,000 years ago. The Septuagint chronology adds almost 1,400 years, and some scholars make a case for this being the correct timeline.

Ancient pagan cultures and accompanying myths had their origins based upon long eons of time. The writers of the books of the Bible, though they were aware of these myths, chose to record the history of the world in this young world timeframe. The majority of church fathers, such as Basil the Great (329-379 AD), Origen (182-251 AD) and Augustine (354-430 AD) believed the days of creation to be 6 24-hour days and that the world was less than 10,000 years old.

About 300 years ago, this timeline began to come under an indirect attack during the time of the Renaissance in the 1700's and 1800's. There was an increasing resistance to the monarchies in Europe and their use of the church to oppress people. One can argue that the politics of the time were not a proper reflection of the practical application of Biblical principles to government. For that reason, it became fashionable for prominent individuals to question the authority of the Bible and the divine right of kings to rule. Many of the leaders of the American and French revolutions were deists who believed in a god of a distant past who set things in motion but left it alone.

MASORETIC CHRONOLOGY

Age of the Earth	Event	Scripture	Date
0	Creation	Gen. 1:1-31	4004 BC
130	Adam 130 Seth born	Gen. 5:3	3874 BC
235	Seth 105 Enos born	Gen. 5:6	3769 BC
325	Enos 90 Cainan born	Gen. 5:9	3679 BC
395	Cainan 70 Mahalalel born	Gen. 5:12	3609 BC
460	Mahalalel 65 Jared born	Gen. 5:15	3544 BC
622	Jared 163 Enoch born	Gen. 5:18	3382 BC
687	Enoch 65 Methuselah born	Gen. 5:21	3317 BC
874	Methuselah 187 Lamech born	Gen. 5:25	3130 BC
1056	Lamech 182 Noah born	Gen. 5:28	2948 BC
1558	Noah 502 Shem born	Gen. 11:10	2446 BC
1656	Flood when Noah was 600	Gen. 7:6	2348 BC
1658	Shem 100 Arphaxad born	Gen. 11:10	2346 BC
1693	Arphaxad 35 Salah born	Gen. 11:12	2311 BC
1723	Salah 30 Eber born	Gen. 11:14	2281 BC
1758	Eber 34 Peleg born	Gen. 11:16	2246 BC
1787	Peleg 30 Reu born	Gen. 11:18	2217 BC
1819	Reu 32 Serug born	Gen. 11:20	2185 BC
1849	Serug 30 Nahor born	Gen. 11:22	2155 BC
1878	Nahor 29 Terah born	Gen. 11:24	2126 BC
2008	Terah 130 Abraham born	Gen. 11:32 Gen. 12:4	1996 BC
2083	Abraham 75 enters Canaan	Gen. 12:4	1921 BC
2513	Exodus from Egypt	Ex. 12:40	1491BC
3420	Last deportations of Jews	Dan. 9:24-27	584 BC
4000	Birth of Christ	Luke 1:5	4 BC

SEPTUAGINT CHRONOLOGY

Age of the Earth	Event	Scripture	Date
0	Creation	Gen. 1:1-31	5326 BC
230	Adam 230 Seth born	Gen. 5:3	5096 BC
435	Seth 205 Enos born	Gen. 5:6	4891 BC
625	Enos 190 Cainan born	Gen. 5:9	4701 BC
795	Cainan 170 Mahalalel born	Gen. 5:12	4531 BC
960	Mahalalel 165 Jared born	Gen. 5:15	4366 BC
1122	Jared 162 Enoch born	Gen. 5:18	4204 BC
1287	Enoch 165 Methuselah born	Gen. 5:21	4039 BC
1474	Methuselah 187 Lamech born	Gen. 5:25	3852 BC
1656	Lamech 182 Noah born	Gen. 5:28	3670 BC
2158	Noah 502 Shem born	Gen. 11:10	3168 BC
2256	Flood when Noah was 600	Gen. 7:6	3070 BC
2258	Shem 100 Arphaxad born	Gen. 11:10	3068 BC
2393	Arphaxad 135 Kainan born	Gen. 11:12	2933 BC
2523	Kainan 130 Salah born	Luke 3:36	2803 BC
2653	Salah 130 Eber born	Gen. 11:14	2673 BC
2787	Eber 134 Peleg born	Gen. 11:16	2539 BC
2917	Peleg 130 Reu born	Gen. 11:18	2409 BC
3049	Reu 132 Serug born	Gen. 11:20	2277 BC
3179	Serug 130 Nahor born	Gen. 11:22	2147 BC
3258	Nahor 79 Terah born	Gen. 11:24	2068 BC
3388	Terah 130 Abraham born	Gen. 11:32 Gen. 12:4	1938 BC
3463	Abraham 75 enters Canaan	Gen. 12:4	1863 BC
3893	Exodus from Egypt	Ex. 12:40	1433 BC

| 4869 | Last deportations of Jews | Dan. 9:24-27 | 457 BC |
| 5225 | Birth of Christ | Luke 1:5 | 4 BC |

English geologist William Smith, Scottish geologist James Hutton and later Charles Lyell, wrote volumes about the geology of the world and popularized a doctrine called "uniformitarianism". They believed that the geological strata represented accumulation of sediments over eons of time and that the key to understanding the past were found by examining the processes we observed in the present. In doing so, these geologists gave an air of respectability to the idea that the timeline in the Bible isn't true, therefore the authority of the Bible is in question, and subsequently, the divine right for kings to rule.

Lyell saw himself as "the spiritual savior of geology, disputing the catastrophe of the great flood of Noah. It was Lyell's work that persuaded Charles Darwin to consider the theory of evolution when he read it aboard the *Beagle*. Lyell and Darwin are buried near each other in Westminster Abbey and ironically it symbolizes the fact that their philosophies became part of the foundation of the church.

Lyell was a British lawyer and geologist. He published *Principles of Geology* in a three volume work, explaining his ideas of uniformitarianism, mapping out an old earth with geologic ages. He believed that the present is the key to the past and that the slow and gradual processes observed today are the same processes that shaped our planet's geology over millions of years.

Lyell didn't believe the Bible as God's Word. He rejected it as an accurate account of earth's history. He actually wrote that he wanted to "free the science of geology from Moses." Lyell was probably best described as a deist. He believed in a distant God who did not intervene in the Universe. Without a God, everything had to be explained naturally, and uniformitarianism gave Lyell a way to explain the world's geology without appealing to God or His revelation to us in the Bible. Church leaders, rather than question

these ideas, came up with the day-age theory and gap theories where they attempted to incorporate long ages into the Biblical timeline. My great-great grandfather's Willson's Fifth Reader, published in 1861, describes both theories as possible explanations for the geological ages introduced by Lyell.

My colleague Dr. Jerry Bergman documents in his many books the consequences of the church's embrace of the uniformitarian doctrine and its incorporation into Christian theology. He shows that Darwinism led to racism, Nazism, communism, materialism, eugenics, abortion, and all sorts of other philosophies that ultimately caused the deaths of millions of people and justified dictatorships and evil governments, some of which were much more oppressive than the monarchies they replaced.

Initially, evolutionary theories were based upon fossil correlation and estimates on how long it would take for sediments to accumulate based upon the speed at which they accumulate in the present. Without regard to the flood of Noah, they came up with hundreds of thousands to millions of years. Later, they employed radioisotope dating which appears to give them the numbers they want. They also interpreted hundreds of thousands of layers of sediments in rocks, "varves", to be annual deposits laid down slowly over millions of years.

In 1841, Lyell visited Niagara Falls. The Falls were difficult to get to then. His research set out to determine the approximate age of the gorge that was excavated by the Niagara River. If you visit the falls today, you can see exactly what Lyell saw.

The gorge cuts through an escarpment and extends about seven miles from the Falls down to Queenston, Ontario. The walls of the gorge are 200-300 feet high and are limestone on top, and shale beneath. Limestone is hard, but the shale underneath is soft. Large chunks of limestone break off and fall into the gorge. Cracks in the limestone fill with water, freeze and weakens the limestone.

A long-time resident told Mr. Blackwell, the son of an eminent geologist, that the Falls had receded about 150 feet during the 40 years he had lived there. That was more than 3 feet a year. He used this information to calculate the age of Niagara Falls to be 35,000 years. Most people blindly accepted his estimate, understanding how water erodes rock. But his age estimate was based upon the assumptions that the rate of erosion was constant, that the recession rate was accurate and that there was no Biblical flood or catastrophic runoff. Instead of critically evaluating these claims, the church capitulated to these long ages, though they contradicted the Bible. But Lyell's conclusions were based on a number of logical fallacies. Because of that, people began to doubt the Bible, assuming its chronology was not reliable in the light of what Lyell had claimed. Lyell did not consider that erosion would have been much faster in the narrow gorge and that Father Hennepin's 1677 drawings of the falls contradicted his ideas.

CHAPTER 5

Why Creation Geology is Important

The Bible is God's Word. We can believe that, or we can rely on the words of men who weren't there and are unable to observe what happened in real time and believe something else. If the Bible is true, it then brings to life ancient history. The gospels of Matthew and Luke record the genealogy from Adam to Jesus Christ, naming all of the generations of individuals. These records match the names listed in the book of Genesis, so that means that the disciples of Jesus took Genesis seriously, as did Jesus himself, who said,

Do not think that I will accuse you to the Father: there is one that accuses you, even Moses, in whom you trust.
For had ye believed Moses, ye would have believed me: for he wrote of me.
But if ye believe not his writings, how shall ye believe my words? (John 5:45-47)

The Jews took Moses as straightforward history and took great pains to preserve his words. If you do a word search on Moses in the New Testament, you will see that Jesus quotes him extensively, and that the doctrine Jesus taught was inextricably tied in with his writings and the history of the world from the very beginning.

If Jesus quoted Moses and believed his words, but if Moses was wrong when he wrote Genesis, it seriously undermines the entire foundation of Christianity. There is a lot at stake in getting this right, and it starts with answering the arguments given by those who use fossils and strata to discredit the Bible.

Indeed, the first chapter of John declares that Jesus was there at the beginning and that He was the creator. Every major tenet of the Christian faith originates in Genesis. God when He created man, he created them male and female, and that is the foundation of marriage. It is the origin of wearing clothes, why there are rules, what is right and wrong, and why we're sinners.

It matters deeply what you believe about Genesis. If you believe that there was death before Adam with millions of years of red in tooth and claw, you negate the meaning of the gospel and the reason why Jesus came to save us from our sin. By one man, Adam, sin came into the world and subsequently death. The whole creation groans as a result. That is why if you are a theologian, and you incorporate the millions of years of evolutionary development in the creation story, ultimately the work Jesus Christ did on the cross to redeem mankind from sin is rendered meaningless.

If you are a student, you need to handle this very carefully. If your teacher is an evolutionist, it is likely you will receive test questions that will have answer selections that are opposite of what you understand to be true. In order to get a good grade, what you need to do is to select the answers the teacher expects. It is not the time to try to convert the teacher. That usually doesn't end well. You will get an education in what the evolutionary point of view is, and you can use that to evaluate what you believe to be true and test it. If you're expected to do research and write a paper, that can get tricky. Make sure you parrot back what the teacher expects so that he or she knows that you have learned the material. Use your best judgment as to when it is appropriate to challenge the teacher. That is a matter of prayer.

CHAPTER 6

Where People Get Stuck

The Conventional Geologic Column

MYA	ERAS	PERIODS	EPOCHS	NOTES
	Cenozoic	Quaternary	Recent	
1			Pleistocene	Man
13		Tertiary	Pliocene	
25			Miocene	
36			Oligocene	
58			Eocene	
63			Paleocene	Mammals
135	Mesozoic	Cretaceous		
181		Jurassic		Reptiles
230		Triassic		
280	Paleozoic	Permian		
310		Pennsylvanian		
345		Mississippian		Amphibians
405		Devonian		
425		Silurian		
500		Ordovician		
600		Cambrian		Shellfish
4500	Precambrian			Algae

Geology textbooks will map out the history of the world using the "geologic column" identifying layers of rock strata and giving them names of eras, periods, and epochs. They assign ages to these based upon the fossils they find in each of them, but they also date the fossils based upon the strata where they are found. The trouble is that very few places in the world exhibit this sequence even partially and though there is a general order found, there are strata missing or even in the wrong order.

Those who read the geology textbooks are under the impression that the dates given are accurately determined by a scientific method, yet very rarely is raw data documented that tells you how they arrived at those dates. If someone hands you an ammonite fossil and tells you that it is between 65 and 240 million years old, there is no physical evidence on it that will help you determine that age other than the fact that it is found in the same rock layer with other similar fossils.

Fossils are found in sedimentary rock. Because sedimentary rock is composed of tiny sand grains or crystals that eroded from someplace else, it is impossible to date sedimentary rocks using radioisotopes. Now it may be possible to date a fossil itself using carbon-14 dating, but if that fossil is found in a layer geologists think is millions of years old, very rarely will they use carbon-14 to date the fossil. Carbon-14 has too short of a half-life (5730 years) to yield a date into the millions of years. If they did, they would be embarrassed to find that radiogenic carbon-14 still exists in the sample and the date is far too young to fit their timeframe. So they would dismiss it as contamination, error or simply not publish it. The only kind of rocks where scientists apply radioisotope dating are igneous and metamorphic.

Radioisotope dates are dependent upon assumptions of uniformitarianism. I like to call it "uninformeditarianism" because the physical evidence does not justify that it is a reasonable assumption. Rocks are not clocks. A clock needs to be accurate, sensitive and be based upon a process that is continuous and smooth. But since we do not have the mechanism to go back in time and observe the process over millions of years, we cannot verify the accuracy or validity of the method. To yield a date, a laboratory will crush the rock and analyze the quantity of the parent and daughter elements found in the present. Afterward, there is no means to cross-check the results since the process destroys the sample.

Assumption 1: The rate of decay is constant and hasn't changed over the course of the life of the rock.

Assumption 2: We know the initial physical makeup of the rock and the quantities of radioisotope elements when it was formed.

Assumption 3: The rock has not been contaminated by the introduction of the daughter element.

Assumption 4: The parent element has not been leached out of the sample.

Assumption 5: All of the measured quantity of the daughter element came from radioisotope decay from the parent element.

Assumption 6: Other physical processes such as differential solubility in water are not responsible for altering the ratio of isotopes.

Besides carbon-14 dating, common methods used to estimate the age of rocks include:

- Uranium-Lead,

- Potassium-Argon,

- Rubidium-Strontium,

- Samarium-Neodymium.

These are all vulnerable to the above assumptions being incorrect.

For those who wish to explore the radiometric dating methods in depth, the Institute for Creation Research led a 5-year study of these methods in cooperation with several other creationist organizations and the results were published in a two volume set under the title Radioisotopes and the Age of the Earth (RATE).[2]

The conclusion of the study was that the first assumption that the rate of decay is constant over time is most likely false. Another finding was that carbon-14 dating is the creationist's friend, yielding measurable quantities in coal, diamonds, and limestone which are assumed to be too old to contain any residue carbon-14.

Those unwilling to challenge the fossil sequence in the geologic column and radioisotope dating get stuck here. That is the beginning of compromise.

CHAPTER 7

Exploring the Field

In the following chapters, we will describe various destinations where you can examine the geological features of the world yourself. I will tell you how to get there, what to look for, and who you might consult at the site.

Usually there is a visitor's center at the various parks, and you can obtain literature that will give the standard evolutionary perspective explaining how geologists interpret what you are looking at. They are looking at the same evidence a creationist sees, but they are operating under different assumptions. Don't allow this to intimidate you but learn how to separate the actual physical evidence from its interpretation.

Some of these sites will require you to do a lot of hiking and exploring. If you do that, research the hiking conditions that the park service publishes and consult with the park ranger. Bring plenty of water and food. If you are doing an extended hike, make sure that you are wearing the appropriate shoes, clothing, sunblock, insect repellent and hat.

Carry a map, GPS, or use walkie-talkies. A geologist's rock pick or hammer can be used to gather samples, usually fossils. Stay on the trails, and keep in mind some of the wildlife may be dangerous. If you ride by mule or horseback, they may on occasion try to get you to dismount by getting too close to the edge of the cliff or rub up against the side of a canyon. Water shoes are recommended if you are wading in a stream, like in the Narrows of Zion canyon. It is very helpful to go with a guide, and in some places there are creationist guides available.

A creationist will look for scientific evidence of catastrophe. Fossils are evidence for quick burial. If you see a dead skunk on the side of the road, It is not long before the carcass of that animal decays

and erodes away, even the bones. Pay attention to the fine sedimentary layers in the rocks and how they are deposited. In some cases, they may not be in the order evolutionists expect. Look for evidence of rock movement. This would include, faults, slickensides, breccia, gouge, conglomerates, and scraping marks.

You will often encounter evolutionary explanations for what you see, and in most cases what the evolutionist sees is the same thing that you see. We have the same evidence. It is a matter of interpretation of that evidence. That being the case, you can ask yourself if geology also exhibits evidence that it took place rapidly rather than over millions of years.

An example of what to look for are the size of the world's river deltas. The deltas of the world's major rivers do not contain enough sediment for millions of years of erosion. If you look at the major watersheds of North America, the deltas of the Mississippi, St. Lawrence, Colorado, and Columbia River watersheds show very little sediment. Except for the bird's foot delta of the Mississippi, all of them are estuaries where the flow of the river exits into a wide basin.

The rate of sediment flowing in the Mississippi River is an average 436,000 tons a day. Low is about 200,000 tons, high 1.5 million tons. That's an average of 160 million tons a year. Over a million years, it would be 160 trillion tons. A ton of sand represents 22 cubic feet. So the volume of sediment in a million years would be 240,000 cubic miles.

Where did all of this go?

In 4,000 years it would be 95 cubic miles of sediment. That corresponds with Biblical predictions.

Let's look at other river deltas. The Nile delta is 10,000 square miles. The Mississippi Delta is 5,000 square miles.

The Ganges/Brahmaputra river delta is the largest in the world, drains the Himalayas at 41,000 square miles.

The sediment discharged from the Mississippi River in a million years would be enough to cover the surface area of the earth at more than 1,000 miles. This is powerful evidence that the geology of the earth does not correspond to deep time.

CHAPTER 8

Red Rock Canyon, Nevada

Red Rock Canyon National Conservation Area is located 25 miles west of Las Vegas, Nevada. You can take Interstate 215 to route 159 west, which is West Charleston Avenue. If you are interested in taking photographs, I recommend starting out early in the morning when the mountains are illuminated by the sunrise.

As you enter the park, there is a spot called Calico Basin where there is a great view of the strata, and a parking area where several trails start.

At the visitor center, there is a display that gives the evolutionary geologic explanation of the area. The strata are out of order with Cambrian on top, then Jurassic, Triassic, and Permian. You can see this at Calico Basin.

The gray rock on top is Turtlehead mountain, Cambrian strata and the red Jurassic is underneath. From the Visitor's center, take the scenic drive to the Calico Hills. To the west you can see Bridge Mountain, Juniper Peak, and the Rainbow Mountain wilderness.

There are various trails you can take through the red sandstone rocks at Calico Hills, and you will notice that they alternate between red and white sandstone. This is apparent at the Sandstone Quarry, which is a few stops ahead.

Of particular interest is a stop a little further on. White Rock Road takes you to the trailhead parking lot. There are two trails, one that takes you around the White Rock Hills, and the other to the Keystone Thrust.

Take the Keystone Thrust trail to your right and climb the mountain. Over the crest, you will approach an area where the gray rock changes to red rock. This spot is a geological "window" where the red rock in the valley is exposed, and the surrounding mountains are "older" than the red rock.

Turtlehead Mountain and Sandstone Quarry

Western Peaks

Calico Basin

Breccia at contact

According to the narrative you find in the literature and in the displays at the visitor's center, the gray Spring Mountains, which include Mt. Charleston to the north moved forty to sixty miles from the west to this position. There indeed is a fault here, and the White Rock Hills protrudes above. But the Keystone Thrust contact exhibits about a foot of breccia (broken up rock) in between, indicating movement, but I would argue that if a mountain moved forty miles it would leave a lot more evidence for that movement.

You can also see the red rock layering in the mountains to the west. Petroglyphs can be found in Lost Creek Canyon. Ice Box Canyon is continuously in shadow and can get quite cold even in the summer heat.

The Pine Creek Trailhead is the last stop on the scenic drive before it intersects route 159. Each of these trails are worth exploring and are a photographer's paradise.

I invite you explore the area also by using Google Earth. Without a doubt, this part of Nevada exhibits evidence for vast upheaval but the biggest curiosity here is that the Cambrian, Jurassic, Triassic, Permian out of order strata is also found 75 miles to the northeast in Valley of Fire. Either two separate overthrusts formed the same out of order sequence or the same event formed both. We are left with a geological mystery. I argue that it is much easier to explain this sequence from a Biblical flood worldview.

I might add that when you visit this area, Las Vegas isn't exactly the most family friendly spot. Avoid the strip and find a family hotel from where you can visit Red Rock Canyon, Valley of Fire, and a little further, Zion National Park.

Consider the evolutionary explanation given on the placards at the visitor's center for the Keystone Thrust:

"THE BIG SQUEEZE. The breakup of Pangea forever changed the face of southern Nevada. As the smaller continental masses started moving away from each other, conditions went from relative geologic calm to intense compression. Collision with the vast crustal area under the Pacific Ocean caused

Bonanza King Thrust

Bonanza King from Below

Gray Cambrian Rock on Top of Red Jurassic Sandstone

the birth of Cascade-like volcanoes and the rise of a great mountain chain through a series of enormous, horizontally oriented fractures called thrust faults between 70-150 million years ago. Like a pile of snow in front of some huge plow, southern Nevada bulged upwards as one immense sheet of limestone or sandstone was stacked up on top of another. One of Red Rock's scenic and geologic wonders, the Keystone Thrust (or Thrust Fault), is one of many such faults found throughout the area. Although hard to estimate the exact elevation of this now-eroded mountain chain, it is though the landscape resembled the eastern side of the Andes Mountains 75 million years ago."

Keep in mind that creationists agree with this scenario, only speeded up. In fact, it really only makes sense in light of the momentum caused by rapid continental movement versus creeping slowly inch by inch over millions of years.

Another sign says this:

"AN ARMY OF CATERPILLARS" 15-20 million years ago geologic conditions in the southwestern US changed once again. Instead of colliding, the two plates now began to bump alongside each other laterally. As they did, pressure holding up the mountains eased and the once continuous "North American Andes" were dissected into numerous smaller parallel ranges, that one geologist quipped looked like '… an army of giant caterpillars marching north from Mexico.' Called 'Basin and Range extension," this event exposed long buried rock layers and subjected them to rapid erosion into cliffs and peaks, creating the modern Red Rock landscape.

CHAPTER 9

Valley of Fire, Nevada

About 45 miles northeast of downtown Las Vegas, Nevada on I-15 is the Valley of Fire State Park. The Valley of Fire Highway leading into the park crosses a number of washes, and the road dips, so take it easy when driving. Soon you will ascend the Muddy Mountains. Take note of these. They are Cambrian rocks on the evolutionary timescale. Soon you will encounter a spot where the road turns to the right then left and descends into the valley. There is a place to park here before you go down the hill. Check out the mountain range and the sedimentary rock that composes it.

According to the displays at the visitor's center, these mountains moved here from somewhere else to lie on top of the younger red rocks below in the valley. As you go down the hill into the valley you will observe a spot where the gray Cambrian rock abruptly changes to the red Jurassic sandstone. The out of order rock sequence you observe here is the same as what is found at Red Rock Canyon.

I would argue that these sedimentary rocks were deposited by water action in the order displayed here, rather than the evolutionary explanation that the Cambrian Muddy Mountains were laid down first, then much later the Permian, Triassic, and Jurassic, only for the Muddy Mountains to be uplifted and shoved over on top inch by inch. Some creationists believe that could actually have occurred in the conditions during the flood of Noah, and I am not disputing that possibility. But the evolutionary explanation is definitely difficult to believe, especially to have it occur in two different places.

The red Jurassic sandstone formations are fun to explore. They are carved into wonderful, odd shapes and at the Beehives, you can find small arches and windows eroded into the sandstone.

Muddy Mountains

Descending into the Valley

Approaching the Overthrust Contact

Beehives

Window at Beehives

Beehives

Window

Muddy Mountains from Beehives

Overhanging Rock at Beehives

The next stop on the road is Atlatl Rock. The ancestors of Native Americans carved these petroglyphs, and you can find these throughout the park. The atlatl is a throwing knife that was used to hunt prey before they invented bow and arrow. One can only guess the meaning of these petroglyphs.

Next to Atlatl Rock is a marvelous box canyon which is worthy of exploration. Just before you get to Atlatl Rock, there is a camping area.

Atlatl Rock Petroglyphs

Atlatl Rock

Canyon at Atlatl Rock

Arch Rock

Bighorn sheep

This part of a dirt road loops around the rock formations and reconnects to the main road near the Beehives. As you continue beyond Atlatl Rock you come to Arch Rock. There is a campground there, too. Wildlife is abundant and you may be able to see some bighorn sheep wandering around.

After you complete the loop on the dirt road and get back on the paved road, you will encounter a spot where the landscape changes again and you will be able to view some petrified logs. These are found in the Permian strata and are an indication of rapid burial. We will discuss petrification more in depth when we get to the Petrified Forest, but this is a process that takes place rather quickly under certain conditions.

The visitor's center is the next stop, and many beautiful and interesting geological sandstone formations can be viewed there.

Read the conventional geological explanation at the visitor's center.

Notice how they connect the Muddy Mountain Thrust to the Keystone Thrust at Red Rock Canyon. Indeed the whole sequence is identical, and uncanny. It makes you think that the Cambrian strata was deposited as sediments right on top in both places and wonder if thrusting ever took place. It calls into question the dates assigned to the strata. But between Valley of Fire and Red Rock Canyon, the geology is quite different. In the middle is Frenchman Mountain just west of Las Vegas, and the geology has the Grand Canyon strata sequence tipped over on its side, exposing the Precambrian Vishnu Schist at one

Petrified Log

Bighorn Sheep

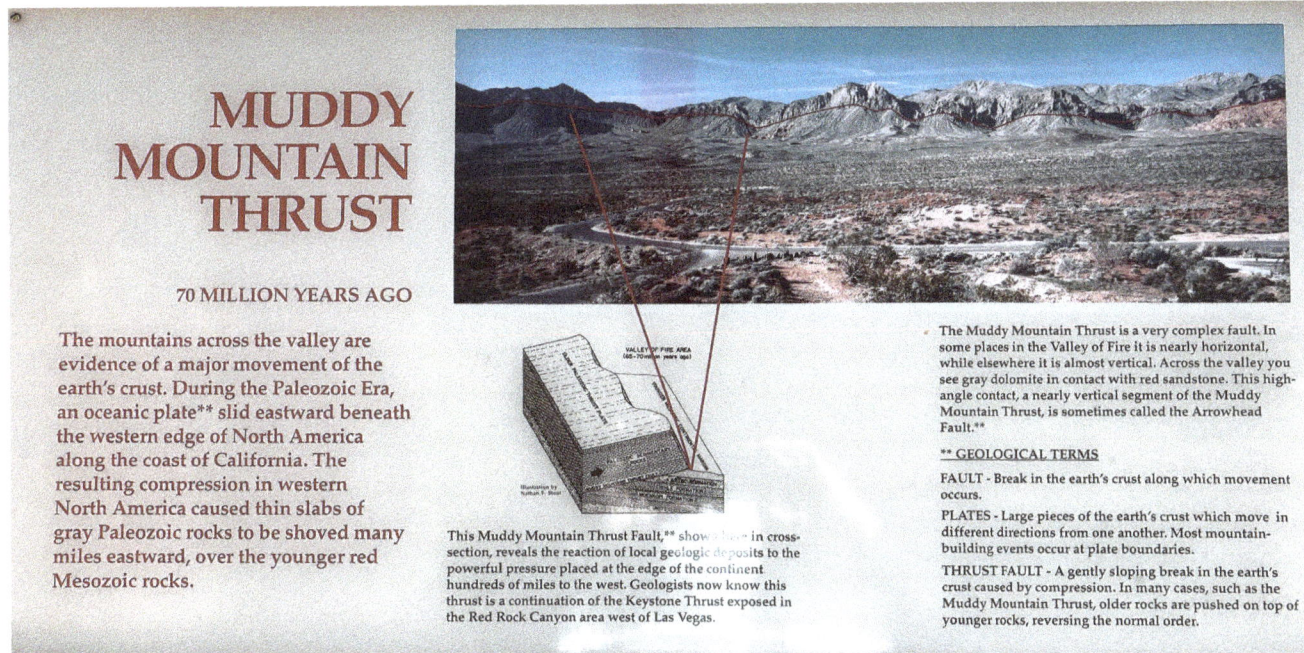

MUDDY MOUNTAIN THRUST

70 MILLION YEARS AGO

The mountains across the valley are evidence of a major movement of the earth's crust. During the Paleozoic Era, an oceanic plate** slid eastward beneath the western edge of North America along the coast of California. The resulting compression in western North America caused thin slabs of gray Paleozoic rocks to be shoved many miles eastward, over the younger red Mesozoic rocks.

This Muddy Mountain Thrust Fault,** show   in cross-section, reveals the reaction of local geologic deposits to the powerful pressure placed at the edge of the continent hundreds of miles to the west. Geologists now know this thrust is a continuation of the Keystone Thrust exposed in the Red Rock Canyon area west of Las Vegas.

The Muddy Mountain Thrust is a very complex fault. In some places in the Valley of Fire it is nearly horizontal, while elsewhere it is almost vertical. Across the valley you see gray dolomite in contact with red sandstone. This high-angle contact, a nearly vertical segment of the Muddy Mountain Thrust, is sometimes called the Arrowhead Fault.**

** GEOLOGICAL TERMS

FAULT - Break in the earth's crust along which movement occurs.

PLATES - Large pieces of the earth's crust which move in different directions from one another. Most mountain-building events occur at plate boundaries.

THRUST FAULT - A gently sloping break in the earth's crust caused by compression. In many cases, such as the Muddy Mountain Thrust, older rocks are pushed on top of younger rocks, reversing the normal order.

end. One wonders at the catastrophic geologic forces at work there. It all adds up to being a real mystery.

There is a short trail at the visitor's center that takes you to the Leaning Rock. The surrounding mountains and weirdly shaped sandstone formations leave you in awe.

Mountains at Visitor's Center

Elephant Rock

Leaning Rock

"Eye of the Eagle"

Petroglyph Canyon

The Valley of Fire Highway continues on toward Overton at Lake Mead, you will pass by the Seven Sisters and Elephant Rock.

But the place you really want to go is up the hill into the red rocks. Just as you reach the crest of the hill, you will come to the Petroglyph Canyon trailhead. This

Petroglyph Canyon

Rainbow Vista

Rainbow Vista

is an amazing trail, and one of the first things you'll encounter is what I call the Eye of the Eagle arch.

What I like about this is that it is difficult to get the perspective. It looks huge in the photograph, but it is about 10 inches high. The Petroglyph trail is also known as the hideout of a Native American outlaw nicknamed "Mouse" who fled authorities and camped here, surviving because of rain water trapped in what is known as "Mouse's tank".

Petroglyphs are found on the canyon walls where the patina is dark gray or black. It is fun for the kids to climb on the rock formations and get a good view.

Beyond the Petroglyph Canyon is the Rainbow Vista and White Domes. It is at the White Domes where visitors often run out of daylight and head back to their hotel.

White Domes

Petroglyphs

White Domes

CHAPTER 10

Zion National Park, Utah

Zion National Park is about a two hour drive northeast of Las Vegas, Nevada, and one of five national parks in southern Utah, all within a day's driving distance from each other. This entire part of Utah is within the Colorado Plateau, and it contains some of the most spectacular scenery in the United States. Some nearby parks are Bryce Canyon, Cedar Breaks National Monument, The Grand Staircase Escalante National Monument, Vermilion Cliffs, Coral Pink Sand Dunes, Kodachrome Basin, Capitol Reef National Park, Goblin Valley State Park, Glen Canyon National Recreation Area, Rainbow Bridge National Monument, Canyonlands National Park, Arches National Park, Natural Bridges National Monument. One could spend a month or two exploring this area and still not run out of eye candy to photograph.

View from Visitor's Center

The Visitor's Center is to your right after you enter the park. From there you can take a shuttle into Zion Canyon, which is closed to traffic.

View from Bumbleberry Inn at Christmas

Springdale, Utah is near the park entrance of Zion, and it is a delightful place to stay. There are plenty of hotels, restaurants, art galleries and museums. The town is surrounded by mountains that display multi-colored layers of rock. The Watchman and Johnson Mountain dominates the landscape on the east and the West Temple, the Altar of Sacrifice and Mt. Kinesava is on your left.

Visitor's Center

At certain times of the year, the Visitor's Center may be crowded, and parking may be at a premium. There is a shuttle from Springdale as well. The Virgin River runs through Zion Canyon creating a riparian environment where it supports enormous cottonwood trees and a diversity of herbaceous plants and grasses. Nearby, saturated wetlands make nice habitat for cattails, willows, aquatic plants, and rushes. In the fall, it is ablaze with color. Camping at the South and Watchman Campgrounds are by reservation only.

The elevation is between 4,000 and 8,000 feet, and if you are used to hiking at sea level, it takes some time to get used to the rarefied air. There are many places in the park where hiking is dangerous

Three Patriarchs

with steep cliffs, loose sand and rocks, and flash floods in the slot canyons. Before embarking on an extended hike, consult the park ranger.

The shuttle has six stops in Zion Canyon, Court of the Patriarchs, Zion Lodge, the Grotto, Weeping Rock, Big Bend, and the Temple of Sinawava. When we visited last, Weeping Rock was closed due to a rock fall on August 17, 2019.

Zion Canyon's rock walls tower 2,000 feet above you. The Three Patriarchs represent Abraham, Isaac, and Jacob. Zion Lodge is maintained by the park service, and from there you can hike to the lower, middle, and upper Emerald pools. At the Grotto, the trailhead to Angels Landing begins along the river and climbs up a series of switchbacks named "Walter's Wiggles".

Trail to Emerald Pools and Angel's Landing

Virgin River

West Rim Trail

West Rim Trail

Angels Landing

Trail to Angel's Landing

Virgin River

Angels Landing

The hike to Angels Landing is a climb of 1,500 feet, and it has long drop-offs. It is not recommended for young children or anyone with a fear of heights.

In 1973, there was a seasonal waterfall at Weeping Rock. 19 people were trapped here and three were injured after a rock fall in 2019.

The shuttle bus does not stop here, and the trail is closed until further notice as of November 2022.

Big Bend is the stop where you can get the best view of the Great White Throne. This imposing monolith rises up with almost vertical cliffs on all sides of the mountain. It is next to Cable Mountain and the Weeping Rock.

The Temple of Sinawava is named in honor of Sinawava, the Paiute's Coyote god or spirit. It is the last stop on the shuttle route. There is a seasonal waterfall that comes over the cliff here, but it is also the trailhead for the Narrows, where the Virgin River flows through a slot canyon about 20 feet wide with vertical walls towering thousands of feet above.

Weeping Rock

Great White Throne

Temple of Sinawava

Virgin River

Path to Narrows

Trail to Narrows

Trail and Great White Throne

Location of Seasonal Waterfall

After the shuttle takes you back to the Visitor's Center, you may take your car to Canyon Junction where you will go up a series of switchbacks to enter the Zion-Mt. Carmel Tunnel. That leads to the East Entrance. The Great Arch is visible from the switchbacks as you approach the tunnel.

Switchbacks

The Great Arch

Switchbacks

Zion-Mt. Carmel Tunnel

Canyon Overlook Trail

Crossbedding

It is important to note the crossbedding and positioning of the layers of sandstone found on the road to the East Entrance. Uniformitarians interpret these to be annual layers of deposits in a shallow sea, where creationists will say that these are all deposited in tidal events that occurred during the flood as Pangea split apart and vast catastrophic forces took place.

Mesa with Crossbedding

Crossbedding

Crossbedding

Checkerboard Mesa

If you take a look at Zion National Park on Google Earth, you will notice that there are quite a number of deep fissures found all running in the same direction north to south as if some giant monster scratched it with its claws. As the uplift of the Colorado Plateau took place, it is possible that these cracks in the earth's crust formed and provided the basis for canyon formation.

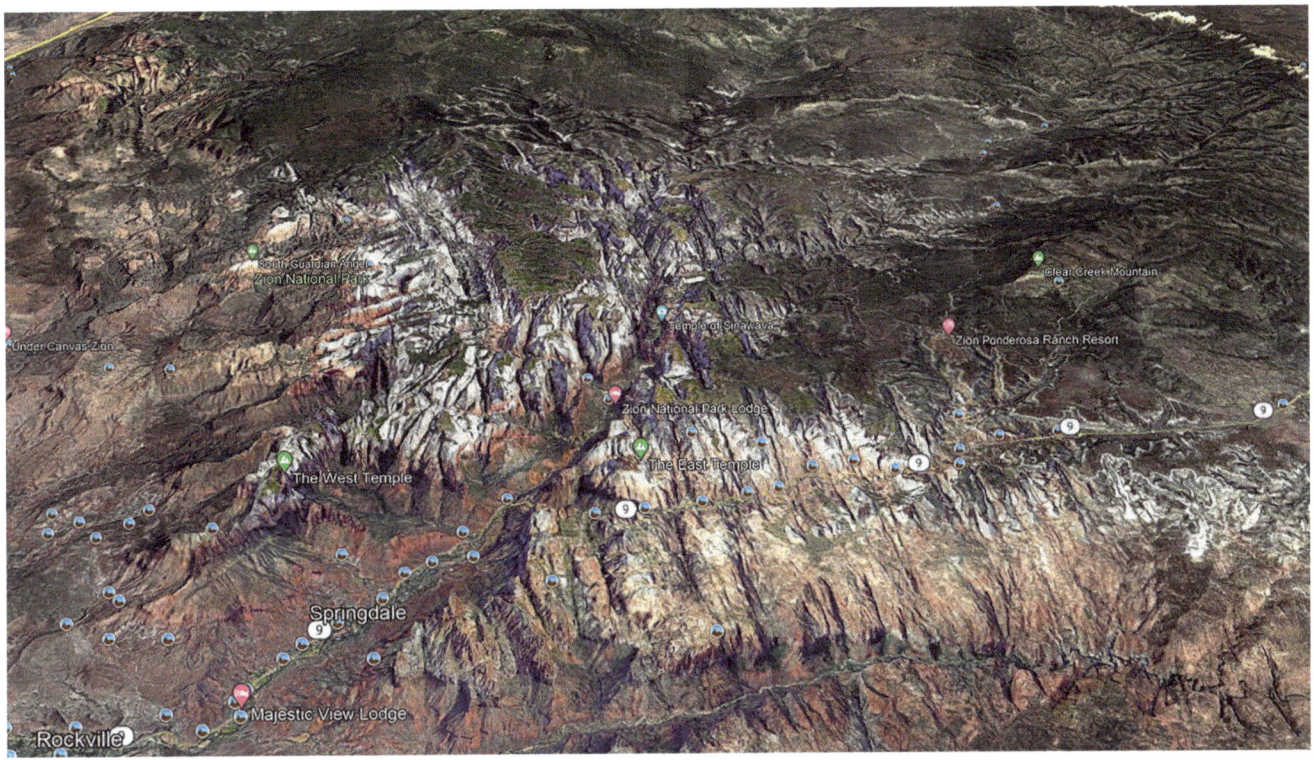

Zion National Park Aerial View

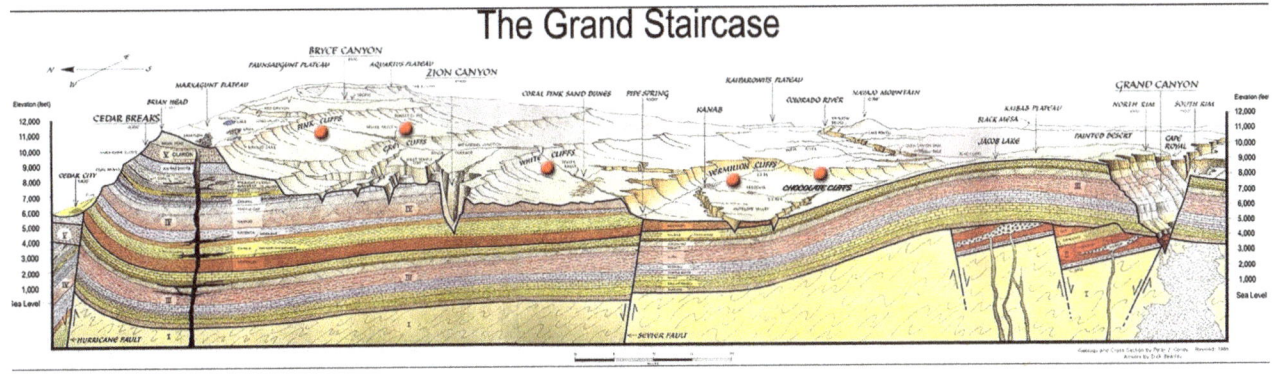

The Grand Staircase

Zion Canyon forms a portion of the Grand Staircase where successive layers of rock make up the Colorado Plateau. The highest elevations are at Cedar Breaks National Monument and Bryce Canyon National Park. We will talk about this in context with the Grand Canyon and other parks in later chapters. Creationists say that all of these layers in the Grand Staircase are the results of many depositions of sediment during the great flood of Noah, and that rapid continental movement would have laid them down all in succession. If there are vast periods of time separating these layers, you would expect to view bioturbation, erosion, and geologic activity between them, but they are laid down on top of each other in succession.

There is another section of Zion National Park that I still haven't visited, and it is the Kolob Canyons section. This is accessible from I-15 at exit 40. The exit is marked "Kolob Canyon" and there is no National Park sign. It is 38 miles away and about a 45 minute drive from Springdale, Utah. If you are spending several days here, it is worth a look. All along I-15, you follow the Hurricane Cliffs where this entire section of the Colorado Plateau was uplifted. The Kolob Canyon drive ascends into the Hurricane Cliffs.

Zion Canyon leaves you with a choice. You can believe that these sediments were deposited in a shallow sea over millions of years, rose up inch by inch and slow processes carved the canyon. Or you can believe that a catastrophic worldwide flood deposited these sediments over 150 days, the continents moved, collided, and were uplifted, and flood runoff created the canyons.

CHAPTER 11

Grand Canyon, Arizona

Creationists have written volumes about the Grand Canyon and have proposed several ideas on how the Flood model explains its formation. There are entire ministries, such as Canyon Ministries, that organize raft trips, rim tours and seminars, and they do an excellent job. I will attempt to supplement the information available from these organizations with some of the observations I gathered from my frequent trips to the canyon.

An excellent resource on the subject is *Grand Canyon: A Different View*.[3] It is a compilation of several creation geologists edited by Tom Vail, who ran Canyon Ministries for many years. For a long time it was offered at the National Park bookstore at the canyon, but the last time I was there, I couldn't find it. Instead, a book by a group of theistic evolutionists, *Grand Canyon: Monument to an Ancient Earth* was in its place. Reviews of that book are on the Answers in Genesis website[4], and I find it interesting that there are a lot of people who find this compromise view satisfying. I don't.

Mary Coulter's Lookout Studio

Lookout Studio

Personally I find that this placard at the Lookout Studio summarizes exactly what people should conclude when they view the canyon. Giving thanks and praise to the Lord should be our first reaction when we approach the edge of the canyon and take in the beauty displayed before us.

The canyon is most often viewed from the South Rim. That is the easiest part to drive to from Flagstaff or Phoenix. It is about 4 hours' drive from Phoenix, Las Vegas, or Grand Canyon West. The Lodge at North Rim as another 4 hour drive. Grand Canyon West is on the Hualapai Reservation, a little over 2 hours from Las Vegas. From Flagstaff, you can take US-180 to AZ-64 to the South Rim. From Williams, you take the AZ-64. The National Park Service operates several lodges at the park, most of them pricey, Maswik Lodge being the least expensive. There are a few hotels in Tusayan just outside the park entrance. For most people who visit the canyon, their first view is from Mather Point.

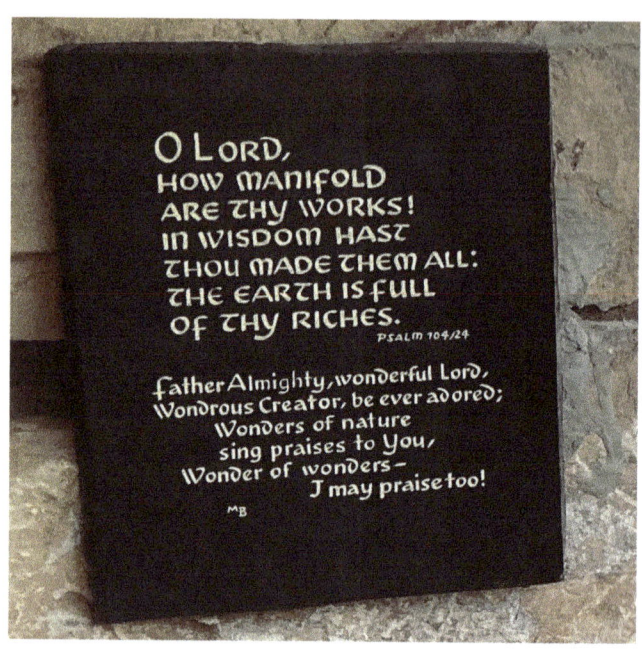

Placard at Lookout Studio

What you notice here is that the horizon and the layers of rock underneath are totally flat and that the scale of deposition is massive. If you notice the white layer just below the top across the canyon, that is the Coconino Sandstone. That layer is distributed across several states and is 50 to 160 feet thick. The conventional explanation for this is that it was a vast

Mather Point

Sunset at Mather Point

desert of sand dunes that hardened 250 million years ago. The rocks you are standing on at Mather Point is Coconino sandstone. Looking across the canyon, there are two other layers on top, the Toroweap and Kaibab. But if you look around a bit, you will find fossil seashells.

Seashells in the Coconino

As I mentioned before, seashells are found throughout the fossil record, and this casts doubt on the frozen sand dune story. It indicates that these layers were water deposited.

Where did these thousands of feet of sediment come from? Did they accumulate slowly over vast periods of time or were they water deposited in successive tidal events as the crustal plates split apart as Pangea broke up?

If you look at the great rivers on our continent, the Mississippi, Missouri, and Ohio rivers, they drain a vast majority of the landscape in the United States. But they are not forming continental sized

strata such as what we find in successive layers at the Grand Canyon. Instead we find a small area at the Mississippi river delta in the Gulf of Mexico. Clearly the uniformitarian model fails us here. What we observe in the present doesn't explain what we see that accumulated in the past.

Bright Angel Canyon

There are many other questions we need to ask our geologist friends. Was the canyon cut by the Colorado River, which seems so tiny in comparison to the huge canyon it flows through? Or is the Colorado River flowing through the canyon after a flood created it? The entire area of the canyon is at high elevation, and it is clear that this section of all of the rock strata was pushed up. In Sedona, the Coconino is 1,000 feet lower than what you find at the canyon. At Frenchman Mountain near Las Vegas, the sequence is tipped over so that the very bottom layer, the Vishnu Schist is exposed.

The flood model explains why there are many layers of accumulation of sediment atop the Great Unconformity, where there are no fossils. It explains why there is a general sequence of fossils that are mostly shellfish and fish throughout, why nautiloids are found fossilized in mudflows pointing in the same direction, why there is a knife-edge contact between each layer without bioturbation, and why some of the layers are interbedded. Megafauna, large land-based creatures, would have been buried

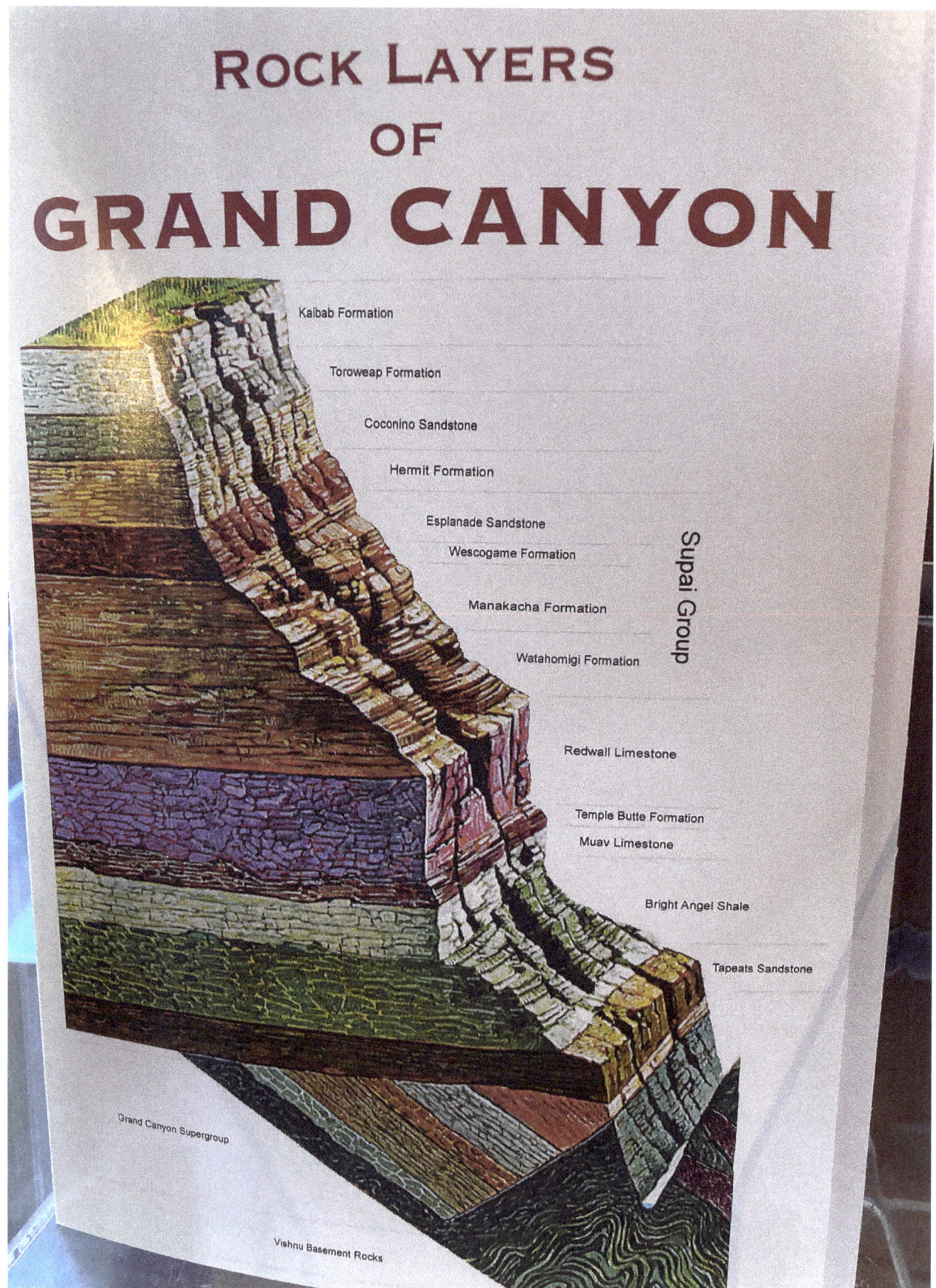

Rock Layers of the Grand Canyon

last, and you would find fewer of them. The Great Unconformity is a series of strata planed off evenly at the contact. As continental plates moved, the entire burial sequence was uplifted by tectonic forces and the entire region was eroded by flood runoff and breached dams to form the canyon.

View of the South Kaibab Trail descending from Ooh-Ahh Point

Bright Angel Canyon in the Distance

In Grand Canyon Village, the Bright Angel trail descends into the canyon, and in the early part of the trail, most of the view is obscured by the surrounding box canyon. A much better trail to take into the canyon is the South Kaibab trail. You take the shuttle to Yaki Point and you can hike 1/3 of the way down to Ooh-Ahh Point. This is a great place to stop and go back. If you plan to hike further than that, you need to be in shape and take lots of water and heed all of the preparations and precautions listed by the National Park service. Rescues can be expensive, and the trail has certain dangers associated with it. The trail crosses the river at the bottom and becomes the North Kaibab trail that climbs all the way to the North Rim. At the bottom, reservations to stay overnight at the Phantom Ranch need to be made well in advance. Camping out in the open isn't a pleasant experience.

There is a passenger train that stops at Grand Canyon Village, and you can also book mule trips into the canyon. Mules can be temperamental, 99% temper and 1% mental. They have learned that if they tiptoe along the edge of the cliff or rub you up against the canyon wall, the possibility that you'll

get off and walk the rest of the way is good. But accidents involving a mule are extremely rare. You can also book a raft trip with Canyon Ministries, and they will give you a creationist perspective on the canyon.

The best Grand Canyon south rim viewpoints are Yavapai and Yaki just east of Mather Point, Pima and Hopi along the Hermits Rest Road, Grandview, Moran, and Desert View along the East Rim Drive. The shuttle takes you to the Hermits Rest Road, and it's a great place to view the canyon at sunrise or sunset. You can on occasion see California condors flying around the cliffs.

The Yavapai Point Geology Center is worth a visit, and it is a good study of the traditional interpretation of the rocks. But I found most interesting the aerial view of the canyon.

Aerial perspective of the Grand Canyon. Note the arrows.

This view from space shows an area in the upper right showing where flood waters spilled into the basin where the Painted Desert is right now, cut a wide area creating two cliffs, one on the right of US-89 to Page, and on the left the Vermillion Cliffs. Some of the side canyons flow perpendicularly or in the opposite direction of the current flow of the river. We might speculate that for a while, canyon erosion and water flow may have been to the east and not the west, creating a large lake.

After you leave the park from Desert View on AZ-64 going east, the environment and elevation changes. There is a steep descent into Navajo country, and you will reach a spot where you can see the Little Colorado River canyon at a Navajo stand.

Little Colorado River Canyon

The amazing feature of this canyon is that it is backwards to what you would expect. It is deepest at the point where it empties into the Grand Canyon but relatively shallow where you encounter it at Cameron. The Cameron Trading Post is historic, worth a stop, and has a great sample of Navajo food in the restaurant. Afterward, make a stop at the Navajo stand near Tuba City where you can see dinosaur and human tracks together. See Chapter 16 for more details about this.

Grand Canyon West can be reached from US-93 taking the Dolan Springs road or from Peach Springs from the east. This is on the Hualapai Reservation

and it's where the skywalk is. The view from there is nice but not quite as spectacular as what you find on the South Rim.

View from Grand Canyon West

Here, the canyon isn't as wide or deep. It is up to you whether you want to do the skywalk, as it is quite pricey and the view from the skywalk is not much different except you can see the canyon under you through the glass. It's a two hour drive from Las Vegas and a 4 hour drive from the South Rim.

The North Rim is also a 4 hour drive from the South Rim, and you need to plan your visit as the only accommodation is at the lodge. It is at a higher elevation and often the North Rim is closed in the spring, snowed in. You will take US-89A across the Navajo Bridge along the Vermillion Cliffs to Jacob Lake and then another hour drive south on AZ-67.

CHAPTER 12

Sedona, Arizona

Sedona, Arizona is nestled among spectacular red rock mountains and formations at the bottom of Oak Creek Canyon which descends along AZ-89A from about 7,000 feet in Flagstaff to 4,200 feet. There is a scenic lookout of the canyon at the top, then descending in a series of switchbacks it straightens out as it follows Oak Creek. It can be rather temperate in Sedona while there is snow in Flagstaff. From Phoenix you can take the AZ-179 exit off I-17. That leads into the village of Oak Creek and takes you into Sedona.

Sedona is home of the Crying Rocks Ministries, and founder Guy Forsythe operates the Gathering Place Christian coffeehouse in town. They publish a monthly creationist newsletter to residents in the town.

Arch in Fay Canyon

Hoodoos

a global scale carrying away the sediments all in one event. The question is where all these rocks went after being eroded.

Fay Canyon

Exploring the surrounding canyons with Guy is a nice experience. The Fay Canyon hike near Bear Mountain is an easy trek for a family and there's a hidden arch along the way.

Formations such as canyons, arches and hoodoos are indications of massive erosion. These are examples from Fay Canyon. Evolutionists insist that these formed over millions of years whereas creationists believe that they are an indication of flood water on

Seismite

This is a photo of a part of the major 500 square meter seismite found in the Schnebly Hill Formation in Boynton and Long Canyons near Sedona, AZ. Over 50 outcrops of seismites have been discovered in the immediate area of about 10 km x 5 km. This seismite is about 6 m in thickness in two contiguous sedimentary beds. Seismites are an indicator of a major earthquake. The Schnebly Hill formation is not found in the Grand Canyon.

The next picture is a parabolic recumbent fold found in the Coconino Formation on Brins Ridge in Sedona, AZ. This sedimentary bed is about 1 m thick but the soft sediment folding extends along the ridge for a distance of about 250 m. The folding is believed to be caused by a change in flow regime of deposition of the bed (now eroded away) above this bed.

Parabolic Recumbent Fold

Lizard Head Recumbent Fold

Water Escape Slits

This is a photo of water escape slits in the Schnebly Hill Formation in Sedona, AZ. The photo shows several thin sedimentary beds, with two of them exhibiting slits. Water escape is caused by a trigger, most likely an earthquake. The shaking causes water and sediment to separate. The water will follow the path of least resistance in its escape.

Usually, the result is a water pipe or injectite that indicates the water moved upward. In this case, there is so much sediment above the area of escape that the path of least resistance is horizontal. Water escape slits are found in many place in the Schnebly Hill Formation over a wide area. They are also found in one location in the Coconino Sandstone Formation and in several locations in the Ft. Apache Limestone (dolomite in the Sedona area) member of the Schnebly Hill Formation. It is difficult for a materialist to explain how there could be so much sediment above the water escape features and there still be so much water in the unlithified sediment.

Lizard Head is the tallest parabolic recumbent fold in the Coconino Formation in Sedona, Arizona. It measures about 3 meters by 1 meter.

Recumbent folds are an indication of change of direction in flow of sediment as it is being deposited.

Another Seismite

Cathedral Rock

Sedona Airport View

This is a photo of a seismite found in Loy Canyon in the Schnebly Hill Formation in Sedona, AZ. Very near this location are several sets of reptile tracks, the only reptile tracks in the Sedona area. The road that goes up Schnebly Hill is very rough and should only be attempted with 4WD vehicles.

These features found at Sedona are powerful indications that they were formed quickly when the sediments were soft before they were hardened. It would be difficult to explain parabolic recumbent folds, seismites and water escape slits with a uniformitarian slow process.

Sedona provides many opportunities for beautiful scenic photography. If you go south of town on AZ-179 to the village of Oak Creek and take Verde Valley Road to the west, follow it until it becomes a dirt road and there will be a parking area where you can hike to Cathedral Rock and Oak Creek. You can also take AZ-89A to Airport Road that takes you to the Sedona Airport Scenic Lookout on the mesa. Look for Coxcomb, Coffee Pot, and Snoopy Rock.

CHAPTER 13

Sunset Crater, Arizona

Flagstaff, Arizona features the highest mountain in Arizona, Humphreys Peak, at 12,637 feet elevation. It is part of a range of mountains known as the San Francisco Peaks. Nearby to the east is Sunset Crater National Monument which features a cinder cone and quite a number of lava flows from an eruption that occurred in historic times circa 1054 AD. Hopi tribal legends recorded this event.

Sunset Crater Cinder Cone

Since this was a historic eruption, it is a good opportunity to test radioisotope dating methods. But when the analysis was done, it yielded potassium-argon dates between 250,000 and 270,000 years. Creationists have documented in the scientific literature a number of examples where they have done radioisotope dating on historic volcanic eruptions that have yielded dates that are far too old.[5] Sunset Crater is one of them, along with Mt. St. Helens, Mt. Lassen, Mt. Etna, and Hawaii.

From US-89 you take route 545 to the National Monument headquarters, where there is a bookstore and museum that is open seasonally. Flagstaff is quite high in elevation, and there is often snow when in Sedona, much lower in elevation, it could be 60 to 70 degrees. At the museum I found an interesting book by Native American Vine Deloria called

Red Earth, White Lies. In it, he talks about geologic events from the past recorded by his ancestors that are different from conventional geology. In his account, he tells of the Three Sisters mountains west of Bend, Oregon that were once all a part of a 15,000 foot mountain, and a legend that the Columbia River Gorge west of Mt. Hood was a giant lava tube with a lava bridge above.

Lava Flows

Lava Flow Trail

Lava Flow at Sunset Crater

As you depart the visitors center, you will notice a giant lava flow next to the road while you are approaching the crater viewing area. There is a trail called the aʻa trailhead where you can explore it. The type of lava called aʻa is clinker like and sharp-edged. Pahoehoe is another kind of lava that is smoother and rope-like.

After you leave Sunset Crater, the road continues through the volcanic area and then connects into Wupatki National Monument. This park features ancestral pueblo ruins, and one of them features the petroglyph known as "Puff the Magic Dragon". Guy Forsythe and David Vonderheide won a lottery for a guided overnight hike to this remote spot and to reach the top of the pueblo, they needed to ascend up a narrow claustrophobic passage.

This is just one of several Native American petroglyphs that resemble dinosaurs. I will discuss more of these in the chapters about the Tuba City tracks and the sauropod dinosaur petroglyphs at Natural Bridges National Monument.

Puff, the Magic Dragon

CHAPTER 14

Meteor Crater, Arizona

I find it interesting that evolutionists spend a great deal of time appealing to a catastrophic event such as the Chicxulub Crater in the Gulf of Mexico to speculate about the demise of the dinosaurs in their uniformitarian theory. There are also some creationists who argue that the Flood of Noah was triggered by impacts from meteorites. I think the idea is intriguing to ponder but certainly not necessary as an explanation.

Indeed, we can cite many examples of impact craters all over the world. One of the largest is in South Africa. The Vredefort crater is 100 km wide and about 40 km deep. In the Canadian Shield in Quebec, the Manicouagan Reservoir fills the perimeter of a large impact crater, the Pingualuit Crater is in the extreme north. The eastern part of the Hudson bay is suspected to have been formed by an impact along with two nearby lakes. The Sudbury, Ontario nickel mines are also an impact crater site.

But the Barringer Meteor Crater near Winslow, Arizona may be the most famous and visible of impact craters on the planet. This is a piece of it found nearby.

The Holsinger meteorite is the largest fragment of the 150 foot meteor that created the Meteor Crater. The site is just off I-40 between Winona and Winslow in the middle of a flat desert.

The Holsinger Meteorite

Meteor Crater

The site is interesting, but it really isn't a thing of beauty. It looks like a big gravel pit. It is difficult to get a perspective of the size of the crater.

Crater Wall Push Up

It is a little less than a mile in diameter and 570 feet deep. The reason creationists have proposed that meteor impacts were involved in the flood is because of the preponderance of craters on the moon and in the solar system. In past models of the flood, more importance was placed upon the collapse of a water vapor canopy, but with the catastrophic plate tectonics model, that explanation can be a factor, but not essential to the model.

A visit to the meteor crater could be coupled to a visit to Walnut canyon National Monument, each taking a half day to explore and experience.

CHAPTER 15

Walnut Canyon, Arizona

Walnut Canyon National Monument is a steep-sided canyon where ancient Native Americans built pueblos. It is a short drive east of Flagstaff on I-40.

At the visitor's center, the trail to the Walnut Canyon Island loop descends to the island with a series of switchbacks. The Island is a part of the canyon that is eroded on all sides, leaving an isthmus connecting it to the canyon wall. It is a moderate trail, and most people should not have trouble with it, but you will need to take the normal precautions about carrying enough water. You are at 7,000 feet elevation and coming back up the trail to the visitor's center is quite a climb of 185 feet, so you are likely to be breathless.

Trail to Island

Pueblo

Pueblo Cave

Canyon Wall Sandstone Crossbedding

What is significant at Walnut Canyon is the crossbedding in the sandstone found in the canyon walls. You will find that it alternates back and forth, favoring the creationist explanation for its formation by alternating successive tidal depositions during the flood. The evolutionary explanation that they were wind-blown dunes that were hardened over millions of years of deposition just isn't quite as satisfying.

CHAPTER 16

The Tuba City Dinosaur and Human Tracks, Arizona

The Cameron Trading Post is at the junction of AZ-64 and US-89 and is worth a stop for genuine Navajo food for lunch and browsing all of the Navajo trinkets, artwork, fossils, and Western wear. North of Cameron about 10 miles is the junction of US-160 which takes you toward Tuba City, Kayenta and Monument Valley. On US-160, you will ascend a ridge to your left, and another about a mile later. As you crest the hill, on your left is a Navajo stand with a sign, Dinosaur Tracks. Make sure you're not speeding, otherwise you'll fly past it and have to turn around a few miles ahead.

This is well worth a stop, and you should allow for a couple of hours here to explore. I have been here at least five times, and each time I see something new and learn a lot. This is Navajo country in the middle of the Painted Desert. I enjoy talking to the Navajo guides who will come out and greet you when you park your car. They will explain the dinosaur tracks that are evident especially just beyond their souvenir stand. The dirt road where the dinosaur tracks are leads to their village Moenave where they grow pistachios. A little further east on US-160 is the two towns of Moenkopi and Tuba City. Your guide will lead you on a tour of the dinosaur tracks and usually he or she will take you on a path to the right. Immediately you will find a whole host of dinosaur tracks, and according to them more often are exposed with successive rainstorms washing away the overburden. As the guides tell their story, it is customary to give them a generous tip.

Of interest to creationists are the occasional human tracks that are found here. The Navajo guides will tell you that their ancestors walked with the dinosaurs, and they will show you the human tracks. The tracks shown below were freshly uncovered and weren't visible on my previous trips to the site.

Dinosaur Tracks

Tracks Next to Navajo Stand

Dinosaur and Human Tracks

Human Tracks

They also told me that they had legends of ptero-dactyls, or thunderbirds, flying around Shiprock in New Mexico near the four corners area. When I visited the tracks in 1999 I found individual human tracks and a sequence of tracks.

Human Track Sequence

These tracks were found about 75 yards from the main track site on an angle to the left. The site also has a lot of in-situ fossils, one of which is a claw of an allosaurus.

Allosaurus Claw

Large Dinosaur Track

Dinosaur tracks

Human Tracks

According to the Navajo guides, the tracks and fossils are found throughout the area, across the road, down the hill and all along the mesa near the village where they live. What this means is that the deposits are laid down first, then the tracks were made and buried in mud, only to be uncovered like here in flash floods or rainstorms. If the trackways are found extending into the mesa and if you can remove overburden to discover more, it means that successive layers are not thousands or millions of years apart, but it is likely that from the lowest level where there are tracks to the highest level, it is the same event that produced these tracks. Since the Navajo claim that their ancestors walked with the dinosaurs, I would place this event to be recent, post-flood.

In some of the spots where the tracks are found, you can see that a few of them are isolated, being harder than the surrounding rock, are left in a track island with the layer underneath around it eroded away. In some cases, fossilized remains of creatures are still embedded in the rock layers, and some of them still show remains of soft tissue, either as impressions or they could be actual soft tissue remains. The one set of dinosaur bones ring hollow when you knock on them, indicating that they are actually bone, not fossil. A few dinosaur skulls are at the site, as well as coprolite, which is fossilized dinosaur dung. Remains of the spines of these creatures, tails, and marine vegetation, all indicate quick burial and preservation in successive catastrophic events where wave after wave of deposition created the layers of mud that turned into rock.

We do not have physical evidence in these sedimentary layers of rock that tell us when these tracks were made and when they were buried. The fact that they have been preserved with a lot of detail would be evidence that they are quite recent and not millions of years old. Carcasses, bone, and soft tissue preserved in the rock also speak to the same story of

Dinosaur Tracks

Creationists also have different ideas about these remains. Some may argue that these were buried in Noah's flood, others would argue that these were buried in a post-flood catastrophe. Dinosaurs would have been on the ark as young, and legends of dragons exist in the stories of people all over the world. So if these tracks and fossils are from a recent event, it would mean that the major geologic events of the flood would have occurred first with continental movement and uplift of the Colorado Plateau.

recent burial. Sedimentary rock does not have properties where we can use radioisotopes to date it, and those have assumptions that are suspect.

Carcass, spine

Reptile carcass

Then over the course of a few hundred years after the flood, creatures such as dinosaurs, mammoths, mastodons, migrated across the polar regions along the land bridge. Because of the warm oceans heated by volcanic activity, those regions were temperate in climate for many years.

Dinosaur Skulls?

We do know that there were many post-flood catastrophes such as volcano activity and earthquakes. The continents were still settling. One explanation of the geology of the Grand Canyon and surrounding area is that a massive lake was built up in basins behind the Colorado Plateau, and that the canyon was formed as a result of drainage of those lakes.

A few fossils presumed to be dinosaur skulls

Turtle Eggs

At the track site, there are quite a number of in-situ fossils such as dinosaur bones, skulls, turtle eggs and carcasses with impressions of soft tissue, skin, and coprolite, which is fossilized dinosaur dung. The fact that some of this is only partially fossilized is an indication that the fossils are relatively recent and recently uncovered. As we have discovered elsewhere, dinosaur tracks and fossils are subject to weathering and erosion, leading to their demise.

Carcass

Marine Vegetation Quickly Fossilized

Marine Vegetation

Coprolite

Dinosaur bones

This set of dinosaur bones ring hollow when you knock on them. They are buried in situ in the sandstone. If you are extra generous in giving your guides a tip and show a lot of interest, they may take you a few miles down a dirt road to see a petroglyph site. This site is fenced and inaccessible except when the guides take you there. Like petroglyphs elsewhere such as at Natural Bridges National Monument, this is a mysterious place.

Petroglyph of Unknown Creature

CHAPTER 17

Petrified Forest, Arizona

Petrified Forest National Park is 115 miles east of Flagstaff, Arizona, and is a little less than 2 hours' drive on I-40. Near the junction of AZ-77 and US-180 outside of Holbrook is Jim Gray's Petrified Wood shop, which, if you are a fossil aficionado, is well worth a visit.

Dinosaur Fossil

Petrified Wood for Sale

Besides the abundance of petrified logs, it is common to find dinosaur fossils here as well. Because dinosaurs are extinct except for endemic species such as Komodo dragons, it is traditional with evolutionary theory to place them between 135 million and 65 million years ago. We of course believe otherwise, and that those dates are dependent upon evolutionary assumptions. You will note that these fossils are found with marine fossils and the huge, uprooted trees in the petrified forest.

There is really no uniformitarian way to view this site. This is part of the Painted Desert, and we have already seen that the dinosaur remains and tracks at Tuba City appear to be recent. Because of the volcanic ash that covers the area just like it is at Tuba City, we can believe that the Petrified Forest was part of the same event.

Petrified Wood

Fossil Ammonite

Aetosaur

Long Logs

Long Logs

Petrified Logs

Petrified Logs

Massive Logs

The petrified logs at the National Park are massive, hundreds of feet long, an indication that the forest that was uprooted and transported here by flood waters was an old growth forest. The trees are of a size that could have been from a pre-flood forest thousands of years old. A sign at a turnoff tells us that the initial stages of petrification may take only decades, but we would contest that it would take millions of years for the quartz crystals to form. How would you experimentally know that? No million year long term studies have been done to prove that. It is a faith statement, an assumption.

Petrified Log Bridge

HOW LONG DOES IT TAKE A TREE TO PETRIFY?

It depends. Environmental conditions, like burial rate and amount of silica in groundwater, affect speed of petrification. The initial stages may take only decades, but it takes millions of years for the silica's molecular changes to result in colorful crystalline quartz.

The logs buried here during the Triassic Period had become solid crystalline quartz by the time *T. rex* walked the land some 135 million years later.

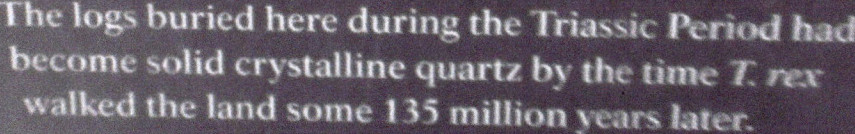

Petrification Explanation

Painted Desert

This scene from the Painted Desert section shows two layers of rock, with the bottom layer cross bedded and planed off. The next layer is deposited directly on top with the contact between the two layers like a knife edge.

CHAPTER 18

Monument Valley

Monument Valley is a Navajo Nation tribal park located on US-163 north of Kayenta, Arizona and is on the Arizona-Utah border. It is also known as Oljato in the Navajo language.

This may be one of the most colorful and interesting geologic areas in the United States. People marvel when they see the Grand Canyon, but these buttes that rise up 1,000 feet above the valley floor are a testament to the removal of all of the surrounding sediment, leaving stand-alone buttes as remnants of the Cedar Mesa to the north.

West Mitten

The West and East Mittens and Merrick Butte

This area takes in the entire area of the Valley of the Gods to the north, the entire San Juan River watershed to the east and Chinle River watershed at Canyon De Chelly to the south. The mesas at the Hopi Indian Reservation rises up from the flat plain to the west. The vertical rock that caps these buttes consist of hard sandstone, and the softer rock below weathers away and causes pillars of the harder sandstone to collapse leaving vertical cliffs.

In a global flood scenario on the scale of what is described in the Bible, imagine that in the early phases of the flood, these sediments were laid down in succession on the ocean floor as the tectonic plates began moving. The sediments would have been much higher than these buttes as the continent rose up while the Rocky Mountains were being uplifted. If you consider that the surrounding plain is flat, these thousand foot tall buttes are all that is left of the surrounding mesas. There's Cedar Mesa to the north. Canyon De Chelly is to the south, Ute Mountain to the east and the Hopi mesas to the west.

Cedar Mesa is just north of Mexican Hat, and you can climb the cliff face on route 261 as you travel to Natural Bridges. Goosenecks State Park displays entrenched meanders on the San Juan River just west of Mexican Hat. Entrenched meanders are a good indication that they were cut rapidly through soft sediment. The Valley of the Gods are buttes carved into the south face of Cedar Mesa and a 17-mile rough dirt road takes you through them. Exploring these sites can give you an idea of the size of the eroded area.

Canyon De Chelly

Goosenecks on the San Juan River

Valley of the Gods (Aunt Jemima)

King on His Throne

The base of the monuments is shale. It is given the name Organ Rock and is 700 feet thick. It is classified by evolutionists as Permian. But throughout the course of the column to the Navajo sandstone on top, all of the layers of rock show marine sediments. Ripple marks, mud cracks, cross-beds, limestone are all evident.

Sisters

Since these formations all are classified as marine sediments, this is consistent with creationist flood geology. We believe that this section of the Colorado Plateau was catastrophically eroded by runoff from the flood. If you can imagine the original deposition from the flood being sediments piled on the ocean floor up to another thousand feet above the highest buttes, with the vertical sandstone cliffs hardened by a cementing agent, then as uplift occurred, the surrounding sediments are eroded away.

Ford Point: West Mitten, Big Indian, Merrick Butte, Castle Rock

Ford Point

Artist's Point

Near Artist's Point

Four Wheel Drive tours of Monument Valley are available at Goulding's Lodge. They are well worth the price. Along the way they will show you petroglyphs and arches such as Moccasin Arch, Spiderweb Arch and Ear of the Wind.

Brigham's Tomb, King on the Throne, Stage Coach, Bear and Rabbit, Castle Rock

Totem Pole and Yei Bi Chei

Stage Coach, Bear and Rabbit, Castle Rock

CHAPTER 19

Natural Bridges, Utah

There are two ways to get to Natural Bridges National Monument. If you are coming from Monument Valley, you will take route 163 to Mexican Hat, then 261 to the Moki Dugway, which is a series of dirt road switchbacks up the cliff face to Cedar Mesa. Once you get to Cedar Mesa, you will enter Bears Ears National Monument which has been newly designated as such. You will connect with Route 95 and then Route 275 takes you to the Natural Bridges National Monument.

This has to be one of the most isolated areas in southern Utah. Aside from the visitor's center, there are no facilities here, so you will need to find accommodations in Blanding or Mexican Hat. If you are coming from the east, you can take route 95 from Blanding. Blanding has an excellent fossil museum.

Natural Bridges is famous for its ruins and petroglyphs along with the three natural bridges. The bridges are named Sipapu, Kachina, and Owachomo. But underneath the middle bridge, Kachina bridge, is the famous sauropod dinosaur petroglyph.

Dinosaur Petroglyph in 1999

The first time we visited the area was in 1999 and we had a lot of trouble finding it. We spent about an hour looking for it and found lots of other petroglyphs on the other side of the arch. Just when we were about to leave, one of the ladies in our group spotted it on the right hand side of the arch wall. The hike down to bridge and the petroglyph site Make sure that you carry some water and travel with a companion. is mildly strenuous. If you are short on time at the National Monument, this is the hike to choose.

Bear's Ears

Panorama at Kachina Bridge

Petroglyphs Traced

Kachina Bridge Trail

Be sure to take lots of water with you and a companion.

At the bottom of the White Canyon, the river there may be a series of mud puddles. There may be spots where sediment has been piled up and it may be tricky to navigate. Be sure to check the weather, as flash floods may be a hazard in the canyon.

White Canyon Floor and Puddles

Dinosaur Petroglyph 2016

When we visited the site in 2016, the canyon had eroded about 15 feet, and the petroglyphs were quite difficult to get to. They were easy to view from the canyon floor in 1999, but in 2016 it took a scramble up the side of the canyon wall in order to get into position to photograph them. The petroglyphs depicts a sauropod dinosaur, and we believe that a second one is also carved to the right of the prominent one. Here we outline them in the photograph, otherwise they are a bit difficult to see.

From what we were told at Tuba City, the ancestors of the Navajo walked with the dinosaurs. That being the case, they observed them enough to make drawings of them. In the Blanding museum, there are dinosaur fossils from various sites in the area. This is a good indication that they existed at least in pre-Columbian times, enough to become part of the Navajo legends. This petroglyph is just one part of an entire series of rock wall carvings.

As you come down the trail from the top of the canyon, the dinosaur petroglyphs shown here are on the right side of the canyon wall underneath the Kachina bridge. There are quite a number of petroglyphs on the opposite side of the canyon and in the side canyons. Again, these were easily visible in 1999, but a very difficult climb to get to in 2016. In 1999, the abundance of these petroglyphs consumed a great deal of our time distracting us while we searched for the famous ones. However, there are a few of them that depict creatures that are unknown, and I will display them here for you to interpret for yourself.

Rhino-like Creature?

Monoclonius?

These petroglyphs raise more questions than they answer. What did the natives of the area 400-600 years ago actually see? What does that tell us of the climate and the culture at that time?

The first bridge in the series of three is called Sipapu Bridge. It is actually a cutoff of an entrenched meander in the White Canyon.

Sipapu Bridge

The temptation at Sipapu Bridge is to take the trail down they canyon wall and explore it, but you have to evaluate what you want to do based upon your timing and where you are staying the night. Both Blanding and Mexican Hat are some distance away, and unless you're camping, there's no town or accommodation nearby. The view of Sipapu Bridge is fabulous from the canyon rim, and I am sure there are more petroglyphs and pueblos at the bottom of this canyon.

The third bridge is called Owachomo Bridge, and it is the largest and most spectacular of the three bridges. The trail here is easy and there is very little effort needed to descend into the canyon and view this bridge.

Kachina Bridge with Tree Branch

Owachomo Bridge

Owachomo Bridge

The park narrative will tell you that this one is the oldest of the three bridges because it is so narrow at the top. I am not sure what scientific data they use to determine that. Erosion rates depend on a number of factors such as the volume of water, the hardness of the rock and how fast the water is moving at the time. If you assume uniformitarian rates of erosion based upon what you observe in the present, that is one interpretation. But having a model for the Biblical flood and its aftermath assuming much greater rates of flood water runoff, each of these arches could be the same age. You can't recreate the past by observing what you find in the present. The flood waters were likely this high.

CHAPTER 20

Montezuma Well, Arizona

On the way north on I-17 to Flagstaff, just beyond Camp Verde are two national monuments, Montezuma Castle and Montezuma Well. Both are just off the freeway and are worth a stop. Montezuma Castle is a spectacular pueblo built into the side of the canyon wall.

Montezuma Castle

Montezuma Well

Montezuma Well is about 15 miles to the north. It is situated right next to Wet Beaver Creek and is a lake with some unusual geology and biology.

The well is almost a perfect circle, like a crater with steep sides surrounding the lake, and it is fed by a spring underneath rich in carbon dioxide. Of interest to evolutionists and creationists are the species of tiny endemic creatures found only here in this lake.

Life in this lake is impossible for most fish, amphibians, and aquatic insects. But there is a amphipod that looks like a shrimp, a tiny snail, a water scorpion, a leech, pondweed, a mud turtle, and an one-celled organism called a diatom. These all have a life cycle where they are interdependent. The diatom is an algae that uses photosynthesis to absorb chemicals from the environment, and the amphipod feeds on it. The water scorpion and the leech feed on

Montezuma Well

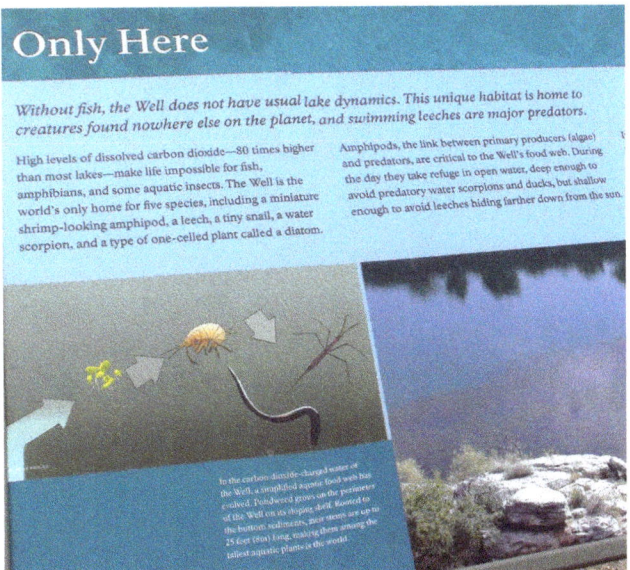

Only Here

Without fish, the Well does not have usual lake dynamics. This unique habitat is home to creatures found nowhere else on the planet, and swimming leeches are major predators.

High levels of dissolved carbon dioxide—80 times higher than most lakes—make life impossible for fish, amphibians, and some aquatic insects. The Well is the world's only home for five species, including a miniature shrimp-looking amphipod, a leech, a tiny snail, a water scorpion, and a type of one-celled plant called a diatom.

Amphipods, the link between primary producers (algae) and predators, are critical to the Well's food web. During the day they take refuge in open water, deep enough to avoid predatory water scorpions and ducks, but shallow enough to avoid leeches hiding farther down from the sun.

In the carbon-dioxide-charged water of the Well, a unique aquatic food web has evolved. Pondweed grows on the perimeter of the Well on its sloping shelf. Rooted to the bottom sediments, their stems are up to 25 feet (8m) long, making them among the tallest aquatic plants in the world.

Only Here — Endemic Creatures

the amphipod. The spring snail requires at least 50 mg per liter of carbon dioxide in the water to survive, ten times what is found in most standing water. The species of pondweed and a Sonoran mud turtle may also be endemic to Montezuma Well. Altogether there may be as many as 12 plant and animal species that are unique to the environment caused by the chemistry of the water in the Well. The water level and temperature are nearly constant, and it has a stable pH because of high alkalinity and the dissolved carbon dioxide.

Pueblo

Along the walls of the cliff face, several pueblos have been built by ancestral indigenous tribes estimated to be from about 900 years ago. These would have been a nice shelter for those who dwelled there, but if they drank the water from the well, it has high concentrations of arsenic, and would have slowly poisoned those who did so. There is a trail that leads from the crater rim down to the spot where the water exits the well through a cave.

Trail to Water's Edge

This is a fairly easy descent, and it will lead to the spot where the water drains out of the Well. There, a sign describes the geologic composition of the Well.

Limestone Cliffs Surrounding Montezuma Well

As you descend into the well, observe the many layers of limestone rock on the cliff face. Though this environment is stable now, it raises the question if this crater was formed as part of a cataclysmic event where these layers were built up over a short time. From what we observe in the present, it is difficult to extrapolate back in the past to reconstruct what might have happened.

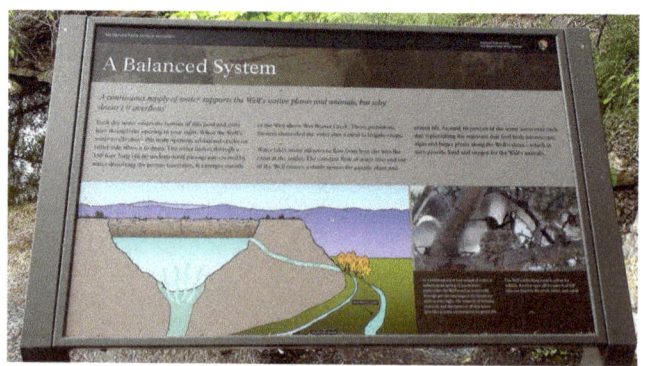

Dynamics of the Montezuma Well

From the top, there is another trail that leads to the other side of the ridge where the water from the well empties into the Wet Beaver Creek. There you can see the outlet from the cave where the water exits the wall of the cliff.

Water Exits the Cave Here

No matter what your point of view is, creation or evolution, explaining Montezuma Well, its chemistry and ecology, is a challenging mystery. The evolutionists assert that the species in the well emerged about 8,000 years ago. However, there is no attempt on their part to explain how these 12 species, totally different, became adapted to this environment.

Montezuma Well Outlet to Wet Beaver Creek

How did these creatures mutually evolve the features needed to survive in this environment and during that evolution gradually adapt? What other creatures or plants did they evolve from? This is not explained. It is simply asserted that it happened by gradual chance mutations.

For the creationist, the mystery still exists, but at least we understand the complexity of the genetic code, and the knowledge that epigenetics play a part in explaining adaptation to environment. We believe that when God created the creatures on the earth, he placed information in the genetic code that coded for all of the environments on the earth. The discovery of epigenetics explains how certain sections of the genetic code are turned on or off by environmental pressures.

The chemistry of life is a dynamic process where DNA codes for RNA, which codes for proteins, and these proteins are involved in all of life's processes, including the manufacture of DNA. It is a dance of non-equilibrium biochemistry with code in four dimensions existing in the structure of DNA, RNA, proteins, where their three-dimensional shapes dynamically change with time and need. Explaining the origin of this process is where evolutionary theory breaks down. Their approach now is to say that this is outside of evolutionary theory. But all life has these processes, they are all complex and to not explain it is a dodge of a critical question.

The existence of the creatures endemic to Montezuma Well speaks of a creator who built into the original genetic code all of the intelligence needed for adaptation to various environments. To assemble new interdependent creatures using random chance mutations by evolution without intelligent design is a faith statement that doesn't have scientific evidence to support it.

CHAPTER 21

Phoenix, Arizona Mountains

The geology of much of the southern Arizona landscape around Phoenix and Tucson reminds you of a flat ocean of sand with icebergs of mountains poking their 1/7 of their mass into the air with the flatlands below. It is worth looking at the geologic maps of the area and trying to interpret what you see through Biblical glasses. Since we cannot reverse time and conduct experiments to simulate what happened, we must look at the geologic evidence we find in the present and from that construct a model of what happened in the past. Uniformitarians believe that you can understand what happened in the past by studying present geologic processes. But it is your starting assumptions that help determine the story of how these mountains were built. They have a story of continents on the move, but at a slow rate over millions of years.

Creationist geologists have built a model of the formation of mountain ranges based upon catastrophic plate tectonics. They would view that a global flood of the proportions described in the Bible would cause a total resurfacing of the earth's crust, and that continents would move rapidly apart from

the mid-Atlantic ridge. There are some features of the mountains surrounding Phoenix that are difficult to explain in the traditional geologic model but fit well into the catastrophic plate tectonics explanation.

If you take Baseline Road in Phoenix to Central Avenue and go south, ultimately it will take you to the entrance of South Mountain Park. South Stephen Mather Dr. passes the ranger station and the connects to West Telegraph Pass Road. This winds up the side of the mountain to a saddle in the mountain where you can view both the north and south sides of the mountain.

The mountain is made up of two different triangular sections oriented WSW to ENE. The westernmost triangle is classified as Precambrian by the uniformitarian geologists and is given a date of 1.7 billion years. The easternmost triangle is classified as Tertiary, and the date they give is 25 million years. But there is a puzzle here. The geologic maps show a section where the Precambrian strata is on top of the Tertiary strata, and the Tertiary strata is exposed underneath all around the Precambrian cap.

Telegraph Pass Looking Northwest Toward Phoenix

Telegraph Pass Looking Southeast Toward Ahwatukee

There is a geologic term for this. It is called a "klippe". Out of order strata like this are common enough for geologists to have a name for it. The opposite of this is called a "window" and this is where there is a valley exposing strata younger than the surrounding mountains. An example of this is found in Cade's Cove in the Smoky Mountains and various other places in the Appalachian Mountains.

In the case of South Mountain, I wonder about the shape of the mountain with two sections almost mirror images of each other in collision about the spot of Telegraph Pass thrusting a section of the Precambrian rock on top of the much younger Tertiary eastern portion. I am not so sure that the uniformitarian explanation that this happened with millimeters of gradual movement is credible. Evolutionary literature doesn't speak of this at all. A feature of this sort needs some momentum to force a thrust fault causing the Precambrian granite to turn to metamorphic gneiss.

Dobbins Lookout

From Telegraph Pass, the road winds toward the summit eventually leading to Dobbins Lookout. Outcrops of volcanic material called dikes appear in this Tertiary section of the mountain. Further along the road, you will reach the Buena Vista Lookout. A trail back up the road from the parking lot will lead you to the Chinese Wall.

Chinese Wall

Papago Park

Papago Park is on the border of Tempe, Scottsdale, and Phoenix and the geology consists of buttes made of breccia, sandstone, fluvial conglomerate, and siltstone. These are all indications of mountain building, flood sediment and extreme faulting. Geologists have a name for this: The Mid-Tertiary Orogeny. "Orogeny" simply means mountain building, and there is no explaining power in this word. They acknowledge that the same mountain building process that created the eastern portion of South Mountain created these buttes as well. Papago Park is the home of the Phoenix Zoo, the Desert Botanical Gardens, and the Hole in the Rock. Many hiking trails traverse the park, and it is a great example of Sonoran desert ecosystem.

West Buttes

There is a parking lot for the West Buttes where there are hiking trails, and opposite that is the entrance to the zoo and the Hole in the Rock.

Butte at Papago Park

North Galvin Parkway traverses the park from north to south. McDowell Road connects Galvin Parkway to Scottsdale and cuts through the buttes. Priest Drive is the exit off the 202 freeway that turns into Galvin Parkway. You can cross the parkway from Curry Drive to Van Buren.

Hole in the Rock

The Hole in the Rock is an easy hike around the back side of the butte where you can scramble up to an arch overlooking the ponds in Papago Park picnic area. As you climb through the hole, there are places to sit in the arch, but venturing further down the side of the butte is dangerous and can be deceiving. The trail goes all around the butte.

Hole in the Rock Approaching the Arch

Papago Park Picnic Area from Hole in the Rock

Camelback Mountain from Hole in the Rock on Back Trail

Camelback Mountain

Camelback Mountain is of the same consistency as the Papago Buttes. It is a distinctive landmark in the heart of the Phoenix area. It is a fairly strenuous scramble to the top. You can climb Camelback Mountain from the Echo Canyon trailhead on McDonald Road.

Camelback Mountain

Hole in the Rock

Again, if you look at the mountains that surround Phoenix and the abundance of flat sediment separating them, you can imagine a cataclysm that shaped all of these mountains during rapid tectonic plate movement during the flood. All of this sediment came from erosion from that event. This includes Piestewa Peak, the McDowell Mountains, Pinnacle Peak, the Estrella Mountains, White Tank Mountains, and San Tan Mountains. These mountains are metamorphized, convoluted, tilted, pushed, and pulled in tectonic plate movements, only to be isolated by the deposition of sediments forming the flat valley of Phoenix. Like one reporter wrote, "It's like going to a car-crushing factory that's been around for a hundred years, and you're asked to go in there and reconstruct all of those vehicles and the makes and models, put them back together and tell us the story about the evolution of cars."[6]

Superstition Mountain

Among all of the mountains in the Phoenix area, the lore surrounding Superstition Mountain concerning Jacob Walsh and the gold from the Lost Dutchman Mine is the most famous. Near the base of that mountain is a museum dedicated to the story, and in the ghost town of Goldfield nearby are further examples of artifacts from that era.

From Phoenix, you take US-60 east to the Idaho Road exit and travel north, then northeast on AZ-88. A few miles down the road you will encounter the Superstition Mountain Lost Dutchman Museum, Goldfield, and the entrance to Lost Dutchman State Park. All three of these are worth exploring, and Lost Dutchman State Park features trails that take you up the mountain.

The Lost Dutchman museum is an old movie set and the museum tells you the history of the Lost Dutchman Mine story. Goldfield is a former gold mine town and has a wild west ghost town theme. But Superstition Mountain itself and the geology surrounding it is from a super volcano that emitted about 2,500 cubic miles of ash and lava. The mountains are composed of volcanic tuff, breccia, dacite, granite, conglomerate, and basalt.

AZ-88 is called the Apache Trail. Beyond Superstition Mountain it takes you into the rough and scenic desert country beyond Apache Lake, Canyon Lake to the wild west town of Tortilla Flat (population 6). There is a rustic campground there next to a wash. Eventually the road turns to dirt and there's one spot near Fish Creek Canyon where it is one lane. We got caught there having to back up into a ditch while a truck with a boat on a trailer was trying to pass. Apache Trail is now closed (2024). Ultimately

you end up at the Theodore Roosevelt Dam. If you are going to do this drive, plan for it to be 15-25 mph, taking all day and staying overnight in Globe or Superior. Sometimes the road is closed.

Superstition Mountain

Superstition Mountain at Goldfield Ghost Town

Goldfield Gold Mine

Typically, you will want to spend quite a bit of time at each of the attractions near Superstition Mountain and you may want to do so on separate visits. The Telescope Lookout trail at Lost Dutchman State Park takes you up to a large boulder on the side of the mountain, and from there you have a great view of the Four Peaks. In the museums you can see amethyst that is mined at the Four Peaks.

Pinnacles at Superstition Mountain

Trail at Lost Dutchman State Park

Lost Dutchman Trail

Big Boulder on Lost Dutchman Trail

Four Peaks

Boyce Thompson Arboretum

If you were taking the Apache Trail to see Tonto National Monument, it is likely that with the slow speed and all of the distracting scenery along the way, you're going to arrive there about the same time the visitor center closes at sunset. It may be better to include a trip to Tonto National Monument at the end of a visit to the Boyce Thompson Arboretum in Superior, Arizona. There is an ancient cliff dwelling at the National Monument and if you are studying Native American pre-Colombian culture, that's what it is all about.

Boyce Thompson Arboretum showcases a collection of plants native to the Sonoran desert, and also has sections dedicated to desert flora from all over the world. Its setting in among the volcanic tuff buttes from the Superstition mountain eruptions give it an other-worldly atmosphere. Whenever I visit the Phoenix area, I try to always make a day to stop here, especially since I have a friend David Oberpriller who gave a Plants of the Bible tour here.

Arboretum Plants

Desert Flowers

Volcanic Tuff

Boyce Thompson Homestead

Volcanic Tuff

Caldera

Boojum Trees

Boojum Tree

To get to Boyce Thompson Arboretum, you take US-60 past Superstition Mountain toward Superior. Just before you get to Superior, there is a turnoff to East Arboretum Way. There is a small sign, and it's easy to miss. That takes you to the entrance of the park. This is a well-maintained botanical garden, and compared to the Desert Botanical Garden at Papago Park, this one has much more acreage and has many trails.

There is a section of the park dedicated to unusual cactus plants, and the boojum trees here come from a spot in Baja California. They are characterized by a fat trunk with short spindly branches. But the

Apache Leap, Superior, Arizona

Volcanic Tuff

Cactus Plants

Bird's Nest

The Potato and the Thumb

Flowering Plants

Variety of Cactus Plants

geology of this area is as much of a feature of the park as the beautiful plants and flowers. This is part of a caldera where a large magma chamber beneath the Superstition Mountains vented, then collapsed, forming a depression. The hot ash and rocks from this volcanic eruption fell back to earth as far away as Globe. The cliff known as Apache Leap just beyond Superior formed in this same eruption. The legend is that in the 1870's, the military ambushed bands of the Apache here and killed many on the spot, but some chose to run their horses over the cliff instead of being killed by the military.

There is an abundance of birds and wildlife in the park. On occasion, arboretum staff and volunteers will host a "scorpion walk" at night where they shine ultraviolet light. The scorpions will fluoresce.

Paths Through Boyce Thompson Geology

Teddy Bear or "Jumping" Cholla

The Teddy Bear Cholla, also known as "jumping cholla", are interesting to look at but need to be avoided at all costs. The spines are barbed and difficult to extract from your flesh.

Multi-columned Cactus

Boojum Trees

Barrel Cactus

Volcanic Tuff in the Caldera

This area between AZ-88 and US-60 could be one of the wildest and inhospitable areas in the world. It is where the Lost Dutchman mine was reported to be.

Certainly the entire Phoenix mountain area shows evidence for extreme cataclysmic events occurring in the past. We believe that it fits the Biblical flood model.

CHAPTER 22

Empire Mountains, Arizona

In 1973, I connected with a group called the Creation Science Research Center in San Diego while I was serving in the Navy. Dr. Henry M. Morris had just formed Christian Heritage College in El Cajon, and the beginnings of the Institute for Creation Research were still a concept about to be implemented. The CSRC was under the direction of Kelly Seagraves and Robert Kofahl, and they had several publications. They had a series of filmstrips. One was called Fossils, Strata and Evolution by John Read, an engineer. He had assembled the research of Dr. Clifford Burdick who was a geologist.

Burdick liked to explore and point out overthrusts, where the fossils and strata are not in the order expected by evolutionists. His filmstrip documented places like Red Rock Canyon, Glacier National Park, Franklin Mountains, Glarus, where the strata were out of order but evidence for movement such as breccia, slickensides, fault gouge and striated stone were not present or minimal. Creation geologists who have come after Burdick have pointed out that in great tectonic upheaval with a lot of water, sections of mountains can move over the surface without leaving a lot of evidence. I'm not so sure about that, and we have very few examples where that has taken place in recent years where this can be tested.

Burdick's claims that out of order strata disproves the geologic column may be taking the argument too far, however we must consider that these conditions exist in geology. Where older rocks rest upon younger rocks, it challenges the assumption that these rocks are really the ages the evolutionists claim. In the creationist Biblical worldview, the rocks can either be pre-flood, flood rocks, or post-flood.

One of Burdick's examples was the Empire Mountains south of Tucson. He showed that strata classified by University of Arizona geologists was Permian on top of Cretaceous. He described the contact between the layers as undulating like a wave and a clean break with no broken up rock between. The rock structure had integrity as if the Cretaceous was laid down first, then the Permian rock was extruded as mud on top.

In 2011, I set out with David Vonderheide to investigate. He had a copy of Burdick's original research, and part of it described an area where he did one of his investigations. Route 83, the South Sonoita Mountain View Highway, hadn't been built when Dr. Burdick did his research, and the old route split off to the left shortly after leaving I-10. Burdick showed several examples of the out of order strata there. We cannot be sure that what we found was what he described, but in a road cut, you could see two distinct layers with wavy contact.

Empire Mountains Contact

Undulating Contact Line at Road Cut

Road Cut

It turns out that the Empire Mountains are a little bit further south, and to find the out of order rocks Burdick described on the filmstrip might have taken us quite a while.

Nevertheless, whenever geologists see rock strata that are not in the order they expect, and publish it anyway, they are raising some questions about how

Bent Strata Contact

they classify the rocks. Appealing to an overthrust, unconformity, or paraconformity to explain the position of rocks is to cavalierly invoke catastrophic processes that might be better explained in a creationist flood model.

Just beyond the Empire Mountains on I-10, south on route 95 near Benson is Kartchner Caverns State Park. These are well worth a visit. These caverns were only discovered about 40 years ago and the State of Arizona has made it into a nice attraction.

When you visit a cave with flowstone, the claim is sometimes made that the stalactite and stalagmite formations take millions of years to form. This doesn't consider a lot of variables, there is no way to test the idea, and evolutionists are beginning to quietly remove those claims, realizing that flowstone is often found underneath bridges and buildings built recently.

CHAPTER 23

Chiricahua Mountains, Arizona

Texas Canyon

If you wish to visit the Chiricahua National Monument, it is in the southeastern part of Arizona almost to the New Mexico border. If you are coming from Phoenix or Tucson, I would combine it with a visit to the wild west mining towns of Tombstone and Bisbee and find some overnight accommodation.

On the way there is a rest area on I-10 at a place called Texas Canyon. Here I fantasize that armies of giants in their anger and fury hurled massive boulders at each other, strewing them all over the landscape. It is worth a stop, and it is loads of fun for kids of all ages clambering all over these huge rocks.

Texas Canyon

Texas Canyon

These rocks are north of the Dragoon Mountains, where south of there the Apache chief Cochise had a stronghold. During the Apache Wars

Texas Canyon

it provided water and shelter for the Chiricahua band for sixty years.

To get to Chiricahua National Monument, you can take US-191 south from I-10 to AZ-181 or from Willcox you take AZ-186. The geology here follows the same pattern you find in the valley in Phoenix, with vast areas of flat desert punctuated with mountains that arise abruptly from the plain. The Chiricahua Mountains are a volcano that is centered around Turkey Creek, much like Superstition Mountain. The remains of the eruption are eroded into thousands of hoodoos. The Bonita Canyon drive to Massai Point can be done in a half-hour but you should allow time to take the hiking trails that are throughout the park.

Here I have decided not to rush to a conclusion about the origin of the geology of this area aside from what was proposed earlier about the Phoenix mountains. Speculating about the past by trying to scientifically interpret what is found in the present is frustrating and wearisome. Instead, I am content to view these amazing formations with a sense of wonder and awe. They have an incredible beauty, and if I try to over-interpret what I see, I may miss an opportunity to worship our Creator and acknowledge that He brings a sense of His power, might, and omnipotence in these expansive views.

Hoodoos

Gnarly Tree

Massai Point

Organ Pipe Rock Formations

Hoodoos

Massai Point

Organ Pipes

Balancing Rocks

Eroded Volcanic Cliff

Grottos

Grottos

CHAPTER 24

Arches National Park, Utah

Park Sign in 1973

Penguins

Park Avenue

This section of eastern Utah would be a fabulous setting for the surreal landscape paintings of Salvador Dali. It is also one of my favorite national parks, and because of its remote location, it takes some planning to visit. It is hundreds of miles from the nearest major airport. The closest airports might be Albuquerque, Phoenix, Denver, or Las Vegas. Since eastern and southern Utah and northern Arizona is the home of six national parks and at least seven national monuments and recreation areas, it would be a good idea to schedule at least two weeks to explore this area.

I first encountered Arches National Park while traveling from Michigan to San Diego on my way to my first Navy duty station in 1973. You could easily get a motel for $6 to $8 at that time. Moab, Utah now has become a mecca for ORV enthusiasts, dirt bikers, and jeep tours. This has hiked the prices of accommodation substantially.

At the visitors center just as you enter the park, there are a series of switchbacks that takes you up the side of a hill, and on the way there's a rock formation called the Penguins. A little further after you crest the hill, there is a parking area for the Park Avenue trail. If you have someone in your group that prefers not to hike, have them take the car to the end of the trail. Then you can enjoy a mile-long hike without having to return to the car.

Douglas B. Sharp

Park Avenue Trail

Park Avenue, Gnarly Tree, The Organ, Courthouse Towers

Looking Back at Park Avenue

Sheep Rock

The Three Gossips

Sheep Rock is the remains of a collapsed arch, and the Three Gossips speaks for itself. Further down the road you encounter the Balanced Rock. When we first saw this in 1973 in the second photo below, a spire called Chip off the Old Block was still standing. That fell in the winter of 1976.

Balanced Rock now and in 1973 in the middle.

This is a high desert area, and the elevation is between 4,085 to 5,653 feet. Since this is sedimentary rock, the water that deposited it must have been above this! This fits in with the catastrophic plate tectonics model where during the flood these sediments were deposited in an ocean, with tidal forces distributing the sediment in successive layers. As the tectonic plates collided, the Rocky Mountains were uplifted along with this entire section of the Colorado Plateau.

Underneath this section of Utah is a layer of salt. The cap rock is much harder, so with the softer rock and salt layer being leached out erosion occurs leaving arches, spires, hoodoos, and weird shapes.

Over 2,000 arches have been discovered in this park. Just beyond the Balanced Rock is a section called the Garden of Eden.

Garden of Eden

Garden of Eden

North and South Window

Turret Arch

   The Windows section features the North and South Window, Turret Arch, Double Arch, and the Parade of Elephants.

Turret Arch

Double Arch

Parade of Elephants

Double Arch

Each of these arches in the Windows section is an easy hike and not difficult to explore from the parking area. In comparison, the Fiery Furnace section should not be explored except with the assistance of a park ranger. This area is a maze, and a lone hiker here can easily be lost or trapped. Anywhere you hike in the park, take lots of water.

Parade of Elephants

Fiery Furnace

Fiery Furnace

La Sal Mountains

The La Sal Mountains in the distance are volcanic laccoliths as are the Henry Mountains to the south, Navajo Mountain to the West, and Ute Mountain near the four corners area.

Delicate Arch Trailhead

Wolfe Ranch

Wolfe Ranch is at the trailhead to the Delicate Arch. The trail to the arch is a mile and a half, and part of the trail is up the side of a slickrock mountain. Good hiking shoes, a hat, and plenty of water is essential for this trek. In places the trail is a little obscure, so it's possible to get lost. The first time I took this trail, I ended up on the wrong side of the bowl shaped depression next to the arch that empties into the canyon below.

Arch in the fin along the trail

The correct trail will take you around a fin with an arch in it, and as you finally get around it, you have a spectacular view of the arch.

Trail to Delicate Arch

Delicate Arch

Canyon Below the Delicate Arch

Canyon on left side of trail to Delicate Arch

The first time I took this hike, I attempted to edge my way along the side of the bowl-shaped canyon where there is only inches of clearance between the fin and sliding down the slickrock into the pit. Underneath the arch, the drop-off is a steep cliff into the canyon below, so watch where you step. Scrambling to the bottom of the bowl can also be tricky. On the other side of the fin you just passed on the trail, there is another canyon on the left hand side. Any of these spots can dangerous, and packing out someone who is injured on the 1 ½ mile trail could be problematic.

In the side canyons, you may be able to spot smaller arches in the canyon walls. Delicate Arch is better left without explanation, retaining your sense of wonder and awe.

Bowl

Bowl

Side Canyon

Arches National Park can be explored in one day, but the timing in which you do it is important. I would start out as close to sunrise as possible. That is when the best light is available to view and photograph Park Avenue, the Windows area, Double Arch and Delicate Arch. Make sure you have enough gas in your tank.

As you return from the Delicate Arch and pass the Fiery Furnace, Skyline Arch can be viewed from the road. In the 1940's this arch was half this size, and a big boulder broke off the left hand side of the arch and collapsed on the canyon floor below. The road ultimately leads to Devil's Garden where there is a campground and a trailhead to the Landscape Arch. It is the 5th largest arch in the world and the largest outside of China. The Devil's Garden trail leads you to Tunnel Arch. Pine Tree Arch and a whole series of fin rocks. Pine Tree Arch actually has a pine tree growing underneath it.

From the trailhead, Landscape Arch is about ¾ of a mile. If you have just done the Delicate Arch hike, you're probably exhausted, but if it is still light out, push on and don't miss this. In 1991 a section of this arch collapsed and a visitor to the park got it on video.

Skyline Arch

Pine Tree Arch

Salt Hill

Landscape Arch

Landscape arch and Gnarly Tree

Landscape Arch

Sunset in Devil's Garden

In historic times, Chip Off the Old Block, Skyline Arch, and Landscape Arch have collapsed or sections of them have. We keep in mind that these rock formations are fragile and still subject to erosion and decay. But we interpret the rocks from a perspective that it was a lot of water over a short period of time rather than a little water over long ages.

Fins

Devil's Garden Fins

I encourage people to view each of these parks on Google Earth, because you can see a perspective from a satellite's view. Looking at Arches, you can see parallel lines of cracks in the earth's surface extending for miles and the extent that the flood waters shaped the landscape.

Fins in Devil's Garden

Fins

Devil's Garden Panorama

CHAPTER 25

Canyonlands and Goblin Valley

Canyonlands National Park is at the confluence of the Colorado River and the Green River. This is a vast area to explore, and if you are going to visit this area, you need to do some planning. The park is in three sections, each accessed from points some distance from each other, and the westernmost section is the most remote and difficult to visit. The north section is the one that probably is the most visited given its proximity to Arches National Park. But it requires an extra day to visit. My first experience in 1973, we were on a time crunch, and we saw Park Avenue, Balanced Rock, Double Arch, the Windows Section and Delicate Arch all in the morning. Then we got as far as the Island in the Sky at Canyonlands, drove on to Capitol Reef and spent the night at Loa. That is an exhausting schedule and not a good way to see any of those parks.

I have been back to Canyonlands National Park twice since, once in 1999 to the southern section called the Needles, and again to the northern section in 2001 with Island in the Sky, Mesa Arch, Upheaval Dome, and Grandview Point. In those two visits, it gave us a cursory perspective of scope of erosion power of the flood. But a lifetime of exploring these canyons could only reveal a portion of the wonders and mysteries of this park. Part of a visit to this area would be to patronize one of the ATV rentals or excursions in Moab.

Most of the park can only be reached by trails and 4WD vehicles. In many respects, the area bounded by Grand Junction on the east, the San Rafael Reef to the northwest, Price, Utah to the north, Glen Canyon to the west, Monument Valley and Canyon De Chelly to the south represents a watershed pummeled by flood erosion that makes the Grand Canyon look like a little ditch. The time and effort it would take to explore just parts of it would be a big project. If you consider the entire Green River, you can start up at Dinosaur National Monument in northern Utah and Fossil Butte National Monument in Wyoming. This is a vast area with many levels of canyons, mesas, and buttes.

North of the Arches National Park entrance is route 313 which takes you into the north section of the park. There is a switchback up the side of a canyon that takes you to a viewpoint of the two buttes called the Monitor and the Merrimac, after the two Civil War Ironclad battleships.

Monitor and Merrimac

The Monitor and the Merrimac

Dead Horse State Park

Dead Horse Point

Dead Horse State Park

Dead Horse Point

Route 313 takes you to Dead Horse State Park which offers an amazing view of the canyon from many viewpoints.

According to legend, cowboys rounded up wild mustangs here and used this as a natural corral. The neck of the point is only 30 yards wide. They fenced it off with branches and brush, chose the horses they liked, and the horses left behind on the waterless point died of thirst.

Dead Horse Point

End of Dead Horse Point

Grandview Point Road leads into Canyonlands National Park and the visitors center at Island in the Sky is the first stop. This area is a high desert and only gets about 10 inches of rain a year.

At Island in the Sky, you can view the dirt roads in the canyon below where you can take an ATV to explore. The paved roads take you to Mesa Arch, Upheaval Dome, and Grandview Point. Make sure you have enough gas, water, and food for the day. There are no facilities in the park.

Island in the Sky

Path to Mesa Arch

On the way to Mesa Arch

View of Cliff

Island in the Sky

Mesa Arch

Island in the Sky

Mesa Arch

Washer Woman View from Mesa Arch

Mesa Arch is a low arch on the edge of the cliff overlooking the canyon, and it has a parking lot with a short trail to the arch. It is easy to climb up and walk on top of the arch, but it is a dizzying view. In the canyon below you can see Washer Woman and Monster Tower, eroded fins of Wingate sandstone. The canyon below the arch is vast. The view extends to the La Sal Mountains into the distance. The canyon at this point is so wide it is difficult to see the other side of it. The Colorado River is way off in the distance. For those who wonder at the forces that carved the Grand Canyon, the scale of this should give you much more to ponder.

Mesa Arch, Washer Woman and Monster Tower

Canyon view from Mesa Arch

Walking on Mesa Arch

On top of Mesa Arch

Grandview Point Road continues on to the south, but Upheaval Dome Road takes you to a circular depression called a syncline. There are two different theories about the origin of this depression. One is that it is the collapse of a salt dome. The other is that it is an impact crater. The latter idea seems to be what is most favored. It exhibits many of the same characteristics of the Meteor Crater in Arizona. Nearby, forming part of the wall of the crater is Whale Rock, which has its own trail where you can scramble to the top.

Side of Upheaval Dome

Upheaval Dome

Grandview Point is a narrow neck of land surrounded by cliffs that descend into the canyon. In the two times I have visited this area, the first time we ran out of time, and the second we almost ran out of gas, so we never really had the opportunity to photograph this spot. If you visit the north section of Canyonlands, make sure you have an ample supply of both.

The south section of Canyonlands is called the Needles district. Route 211 is a turnoff to the west at Church Rock between Monticello and La Sal Junction. This is an entire day's trip, so plan your time accordingly. About ten miles down the road as you descend into the canyon, you will encounter the Newspaper Rock State Historical Monument.

Newspaper Rock

This set of native petroglyphs are interesting in that they exhibit six fingered handprints and footprints. The interpretation of these petroglyphs are a mystery, but we can speculate that polydactyly may have been observed by those who carved these petroglyphs.

The distance from Church rock to the visitor's center at the Needles district is about 35 miles. The descent into the canyon is a pleasant display of eye candy with the red sandstone rocks in the canyon contrasted with the green riparian environment of Indian Creek. This is now part of the Bears Ears National Monument, established in 2016. During our visit, we were warned to beware of flash flooding that can occur in the park, and we encountered threatening ominous dark clouds as we reached the visitor's center. Despite this, we arrived at the Big Spring

Canyon Overlook. There are miles of ORV and hiking trails. Many of the more interesting rock formations can be viewed only by planning an overnight hike or off road adventure.

Oncoming Storm at Canyonlands Visitor's Center
Needles District

Wooden Shoe Arch

The Needles

Butte at the Needles

Crossbedding at Needles District

Weathered Sandstone

Canyon at end of Trail

Needles District

At the Parking Lot

Buttes at Needles District

Needles District

Needles District

Rustic Campground

Canyon in Needles District

Canyon in Needles District

Approaching Storm

The western section of the park takes a great deal of planning to explore. Accessible only by dirt roads from Utah 24, you can plan on a trip of over a hundred miles into the interior of the park. There is a section called the Maze where the canyons are so convoluted that getting stuck there might end up costing you your life. Certainly you will need to consult the ranger station if you want to visit this section.

Just off Route 24, Goblin Valley State Park is worth planning some time for a visit. You can plan for about an hour. This was the site for a scene from the movie Galaxy Quest where a rock monster attacked the spaceship. Indeed it looks like the entire valley is invaded by goblins.

Goblin Valley

Goblin Valley

This is another case where the harder cap rocks are undercut by erosion on the softer rock below. What is amazing is how few of the boulders on top are found collapsed onto the floor of the valley. This formation is other worldly and mysterious. In the distance you can see the San Rafael Reef. The pioneers, from their experience at sea, saw rocky cliffs impeding their progress as "reefs".

Hoodoos

Hoodoos at Goblin Valley

Goblin Valley

Goblin Valley

Goblin Valley

Goblin

Hoodoos

Hoodoos

Goblin

CHAPTER 26
Capitol Reef, Utah

When you approach Capitol Reef National Park, you are unprepared for how remote and desolate this place is. The first time I came here in 1973, we had just hiked in Arches, saw a little bit of Canyonlands and were breezing through on our way to our motel in Loa. We didn't have enough time to linger and appreciate the incredible beauty of this place and only stopped at one spot to hike into a canyon.

Capitol Reef Park Sign

Capitol Dome

Fremont River

Pectols Pyramid

The place that we stopped was at Hickman Bridge, and the trail took me over the top of the hill and into a canyon that was absolutely beautiful in the waning hours of the afternoon. Unfortunately the shadows had overtaken the bridge, and I was not able to get a good photograph.

Canyon on the trail to Hickman Bridge

As nightfall proceeded, we continued on to our motel in Loa. But we were able to get a much better perspective of the park on our visit in 1999.

We had finished our visit at Goblin Valley, it was mid-afternoon, and we were starving. We got some potato chips at a convenience store in Caineville, and that's the only sign of life or civilization in the area. You get a sense of how vulnerable you are if you are stuck here out of gas or water and wonder how early pioneers fared more than a hundred years ago. The "reef" is a giant fold in the earth's crust called the Waterpocket Fold and it extends almost to Glen Canyon to the south. The road Utah 24 follows the Fremont River, and it takes you to the visitor's center on the western boundary of the park. Motel and restaurant facilities are available in the little town of Torrey just beyond the park. When you realize that the nearest cities are a day's drive away, a sense of isolation hits you. The drives from here to Las Vegas, Salt Lake City, Flagstaff, Albuquerque, or Grand Junction are along two-lane mountain winding roads.

Scenic Drive

This is all part of the Colorado Plateau, and the geology interpretation from a creationist point of view works well. The layers of rock are laid down flat like pancakes and the entire region is uplifted intact except for major folds like the Waterpocket Fold here, the San Rafael Swell, and Comb Ridge near Bluff. The fold is a monocline where this section was warped up. Creation geology takes what conventional geologists ascribe to long eons of deposition and speeds it up over the course of Noah's flood. It provides the energy to create all of the layers of rock thousands of feet thick and with the movement of the continental plate and with subduction underneath, a mechanism for uplift. This upheaval isn't explainable from creeping along a millimeter at a time over millions of years.

Just beyond Caineville at the park entrance, the Notom road will take you all the way to Bullfrog Marina on Lake Powell. This follows the entire length of the Waterpocket fold. But just before you get to the visitor's center, Camp Ground Road takes you to the Fruita Campground and the Scenic Drive south. This will take you to Cassidy Arch and the Capitol Gorge. Capitol Gorge is the historic way the pioneers traversed the Waterpocket fold.

Scenic Drive

Cliffs Along Scenic Drive

Capitol Gorge

Capitol Gorge

Big Rock

Capitol Gorge

If you decide to drive to Bryce Canyon from here, a marvelous scenic route is Route 12 south through the Grand Staircase-Escalante National Monument starting from Torrey. Otherwise from Loa, route 24 heads north around a mountain until you connect to route 62 south.

Capitol Dome

Since this area is so remote, it is unreasonable to expect that there would be cell phone service, or that your GPS would work. If you are traveling with more than one vehicle, I would recommend that you carry a walkie-talkie in each vehicle so that you can communicate. Getting lost in this desert wilderness could quickly become dangerous.

Route 12 view of Capitol Reef with Henry Mountains in background

CHAPTER 27

Bryce Canyon, Utah

Bryce Canyon National Park is a high desert and is a divide between the Sevier River that flows north into the Great Basin and the Paria River that flows south connecting to the Colorado River. When geologists talk about the Grand Staircase, they are talking about the layers of strata that begin here on the top.

If you have come from Capitol Reef via Route 12, the Grand Staircase-Escalante National Monument is what you have just passed through. From the west, Route 12 connects to US-89 and that leads through Red Canyon.

Where Zion National Park is massive and viewed from the canyon floor, Bryce Canyon is delicate with expansive viewpoints from the rim. Sunset Point is where most visitors experience their first view of the canyon.

Thor's Hammer

Panorama at Sunset Point

Sunset Point Rim Trail

From that spot, there are hiking trails down into the maze of the canyon. Thor's hammer is a hoodoo that has stood the test of time. It really hasn't changed since the first time I viewed it from here in 1973. I would think after 50 years it would show signs of collapse. One rancher who lived in this area described this place as "a heck of a place to lose a cow". Indeed it would be an easy place to get lost if there were no trails.

Sunset Point

Precarious Edge

Slot Canyon

Stay away from the edge of the cliff. The rocks are unstable and crumbly. You can see that new hoodoos are emerging from the eroded cliff face.

Pinnacles in the Distance

A Slippery Slope to Sure Trouble

Vertical Canyon Walls

Heck of a Place to Lose a Cow

Emerging Hoodoos

Trail into Bryce Canyon

Slot Canyon

Thor's Hammer

Sunset Point

Pinnacles

Park Forest

Closed Trail into Canyon

Grand Staircase

Pinnacles

Slot Canyon

Emerging Hoodoos

Pinnacles

Trail into Canyon

Down the Trail

Warning Sign

Bryce Canyon Vista

Chessmen

Slide into the Abyss

Pink Layers

Vista across the Valley

Ghostly Pinnacles

Castles in the Distance

Trail to Sunrise Point

Trail Below Sunrise Point

Sunrise Point Observation Platform

Sunrise Point

Sunrise Point

Basin of Hoodoos

Be Careful

A Steep Slide into an Abyss

Approaching the Viewpoint

Crumbly Side of Canyon

Scrub Pine Clinging to Side of Cliff

Slot Canyon

Trail View

Canyon View

Gnarly Tree

Root Erosion

Clinging to the Side of the Canyon

Trail to Sunset Point

Eroded Point

Forest Path

Trail Along Canyon

Trail along Canyon

Sunset Point

Inspiration Point Sign

Inspiration Point

Vista at Inspiration Point

Inspiration Point

Inspiration Point Panorama

From the parking lot at Inspiration Point there is a trail that takes you to the upper viewing area. This can be quite a climb to the top and with the high elevation, you will find yourself huffing and puffing.

Over the Edge

Treacherous Slope

Inspiration Point

Erosion Exposing Tree Roots at Cliff Edge

A Forest of Chess Men

Army of Terra Cotta Warriors

Inspiration Point Panorama

Hoodoos

Inspiration Point

Carved by Erosion

Inspiration Point Basin

Bryce Point

View of Grand Staircase

Bryce Point

Bryce Point

Grottos at Bryce Point. Only birds can venture here.

Grand Staircase

Inspiration Point

Bryce Point

Into the Distance

Overlooking the Grand Staircase

Grand Staircase View

Bryce Point Log

Paria View

Standing Up Country

Paria View

Paria View Hoodoos, bottom: Trailhead

    Paria View and Bryce Point are at two opposite sides of a promontory. Other views like Yaki Point are much further down the road, and you need to plan accordingly if you want to see them. They are at higher elevation as well.

Paria View Trailhead

Paria Path

Fairyland at Sunset

The road to Fairyland Trail is near the entrance of the park just before you get to the contact station.

Fairyland Trail and Castle

CHAPTER 28

Cedar Breaks, Utah

From Bryce Canyon, if you go through Red Canyon to get to US-89, you can drive north to Panguitch and then take route 143 south to Cedar Breaks National Monument. This is actually a volcano, and you will climb to over 10,000 feet. Elevation sickness is just one of many hazards here, as it may be at the Grand Canyon and Bryce. Symptoms are headache, nausea, tiredness, confusion, and shortness of breath. The cliffs here are steep and dangerous. Snow is likely here and also depending on the season, the facilities may not be open. No cell phone service either.

Top of the Grand Staircase

Cedar Breaks

Caldera of Cedar Breaks

The view from here is fabulous and it represents the top layers of the Grand Staircase, even above that of Bryce Canyon. Flood deposits from here on top contain oysters, gastropods, and other marine creatures. The fossils represent flood burial. They are not found in a sequence you would find in a geology textbook. The explanation given is that there was once a large lake here the size of Lake Erie, but this does not explain the thousands of feet of sedimentary strata that extend all the way down to the Great Unconformity at the Grand Canyon.

Giant Pit

Top of Grand Staircase

CCC Memorial

Cedar Breaks is a Volcano

Eroded Side of Caldera

Trees Clinging to Side of Cliff

10,000 Feet Elevation

Vast Canyon

Canyon View

CHAPTER 29

Glacier National Park, Montana

Glacier National Park in Montana is the home of some of the most spectacular mountain scenery in the United States. It also presents a giant dilemma for evolutionists. This entire park is upside down. The upper sequence of strata is classified by evolutionists as Precambrian, and underneath is Cretaceous. This is known as the Lewis Overthrust.

The Precambrian is supposed to be the epoch that contains no fossils, or very few. It is considered to be the strata that was deposited before life evolved.

It is also called the Belt Supergroup and is up to 14,000 feet thick. The layer underneath, Cretaceous, however, contains dinosaur fossils.

The area of Glacier National Park in question is 350 miles from north to south and 50 miles wide west to east. It starts in Alberta and at its widest point, the out of sequence strata is exposed all the way around Chief Mountain. The geologic term for it is a *klippe*. Divide Mountain at the eastern end of St. Mary's Lake is also a klippe.

Divide Mountain (in the distance)

Chief Mountain

The trouble is that a section of the Rocky Mountains 350 miles long, 14,000 feet thick does not tend to move anywhere. The forces required to do so would be prohibitive. If the Belt Supergroup was uplifted above the Cretaceous and shoved over 50 miles, the amount of broken rock and evidence for cataclysm would be very evident. Also, as the distance increases, the amount of force needed to push the rocks also increases. Evolutionists, though, propose that the mountains may have slidden down a slope. But considering that these mountains crept along inch by inch in a uniformitarian way over millions of years to get into this position stretches our credibility. This may have been possible during the upheaval of the Flood, given enough water to lubricate the contact so that the mountains could slide. But certainly a possible conclusion is that the mountains were water deposited in this out of order

Chief Mountain from Canada

sequence, and the geologic column itself is brought into question.

Running Eagle Falls is also known as Trick Falls. When the water is low, the flow of the falls comes out of a cave. In the spring when the water is high, the water flows over the top. The trail is at the Two Medicine campground on the east side of the park. Just underneath the falls is the contact of the Lewis Overthrust. Here, geologists have dug out the contact and it is like a knife edge with good integrity of the rock structure above and below.

Running Eagle Falls

Running Eagle Falls (Trick Falls) in 2002

Running Eagle Falls (Trick Falls) in 2021 at peak

Local Wildflowers

Outflow from Trick Falls

Hiking to a Glacial Lake

Two Medicine

Driving to Logan Pass

MacDonald Falls

Trick Falls

Avalanche Lake

Trick Falls

Arête

Upper Macdonald Creek

St. Mary's Lake and Wild Goose Island

Glacier Flow

Beargrass

St. Mary's Lake and Wild Goose Island

Wildflowers

Mountain Vista

Waterfall

Avalanche Falls

Waterfall

St. Mary's Lake

Lake MacDonald

The Garden Wall

Lake MacDonald

Wild Goose Island on St. Mary's Lake

Lake MacDonald

St. Mary's Lake and Wild Goose Island

Entering Waterton-Glacier Park in Alberta

Maskinonge Lake

Dandelion Puffball

Wildflowers

Bonecrusher Rapids

CHAPTER 30

Joggins, Nova Scotia

The Bay of Fundy dividing New Brunswick and Nova Scotia in Canada is the home of the highest tides in the world. They can get as high as 53 feet. The little town of Joggins, Nova Scotia is the home of unusual fossil cliffs on the Cumberland Basin of the Bay of Fundy that are tilted toward the southwest and represent over 18,000 feet of strata. Because they are in such an active tidal basin, they are subject to continuous erosion and the face of the cliffs crumbles, exposing new fossils.

THE FOSSIL CLIFFS OF JOGGINS

"...the place where the delicate herbage of a former world is now transmuted into stone." Abraham Gesner, 1836

The fossil cliffs of Joggins were made famous by the pioneering geological studies of Sir William Logan, Sir William Dawson and Sir Charles Lyell. Sir William Logan founded the Geological Survey of Canada in 1842 and this was the site of his field project in 1843. In 1851, Dawson and Lyell discovered fossil skeletons of amphibians and some of the first reptiles that evolved on earth entombed within the stumps of fossil trees in this area. These cliffs hold an unrivalled record of life in the primeval forests of the Carboniferous Period, some 300 million years ago. The Joggins fossil cliffs are protected under provincial legislation by the Nova Scotia Museum.

It is important to note that Charles Lyell, when he studied this area and assigned it to the Carboniferous period, did not recognize clear evidence of rapid burial in a global flood, and totally ignored the fact that trees and plants buried in this strata sequence extend up through several strata. That being the case, they are the product of one event and not millions of years of continuous deposition.

Furthermore the tilted strata are planed off flat at the top, indicating that after they were laid down, massive flood runoff after uplift washed away the top layers catastrophically. The Cumberland Basin curves around to form the muddy River Hebert, which flows backwards at high tide and reverses at low tide.

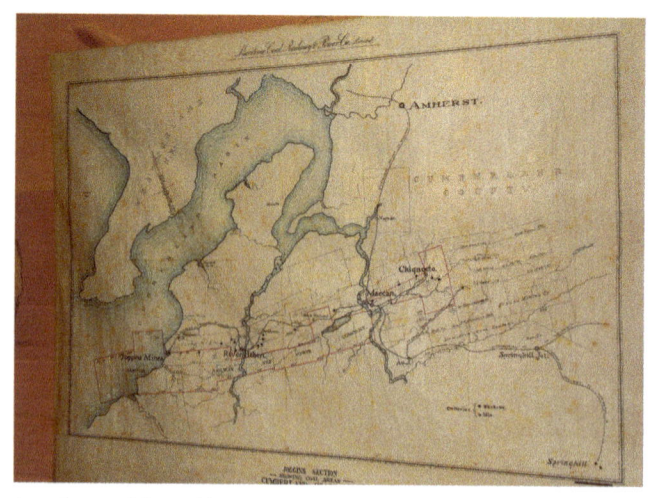

Joggins and River Hebert

Tilted and planed off strata at Joggins, Nova Scotia

Ian Juby and Vance Nelson at low tide

Polystrate fossil at Joggins, Nova Scotia

Polystrate Trees

Fossil Leaf

Getting to Joggins isn't easy. We had to drive from Boston through Maine, New Brunswick and into Nova Scotia. Joining us there was creation scientists Ian Juby, Vance Nelson, and David Vonderheide. Because the tides can be dangerous, we had to time our exploration accordingly so we wouldn't be caught against the cliffs as the tide came in.

What interested Logan, Dawson and Lyell were the fossil amphibians and reptiles found buried in some of the tree stumps. Joggins is the site of coal mines, and 18,000 feet of tilted layers.

Fossil Cliffs

When Lyell and his colleagues examined the layer of coal among the strata found here, their paradigm excluded the possibility of a worldwide flood and they concentrated on the types of fossils found here and not the physical evidence for rapid burial.

Fossil Tree Trunk Extending Through Several Strata

The trees are giant Calamites and Stigmaria. Here in Michigan, Grand Ledge has similar sandstone cliffs eroded in the Grand River. Stigmaria and Calamites fossils are found here as well in the rock layers. These are types of horsetails that grew to giant size before the flood due to a favorable environment that allowed creatures and plants to grow for hundreds of years. Evolutionists refuse to consider that explanation.

Low Tide

Fossil Tree Trunk imprint at Joggins.

Low Tide Exposes Tilted Layers Planed Off

Low Tide

Low Tide at Parrsboro

Low tide at Joggins exposes the jagged edges of the tilted layers of rock extending out as much as a mile. Across the peninsula from Joggins on Mimas Basin is the town of Parrsboro. There you can see the tremendous effects of low tide causing boats that are docked to rest on platforms as the tide goes out.

Low Tide Showing Tilted Layers

CHAPTER 31

Michigan Bedrock Geology

Michigan's geology is dominated by glacial features on the surface, but underneath is a pattern that looks like a basin.

The sedimentary rocks that form the bedrock proceed from the youngest at the center and the circles of layers proceed older in sequence the further you get from the center. Of note are the layers in pink in the center which are classified as Jurassic Red Beds. The yellow Grand River formation is the next youngest, classified as Pennsylvanian. The Ledges are sandstone cliffs in this formation and contain fossil reeds called lycopods and stigmaria, and these can be collected at Lincoln Brick Park. Curiously, these are the same type of fossils found at Joggins, Nova Scotia.

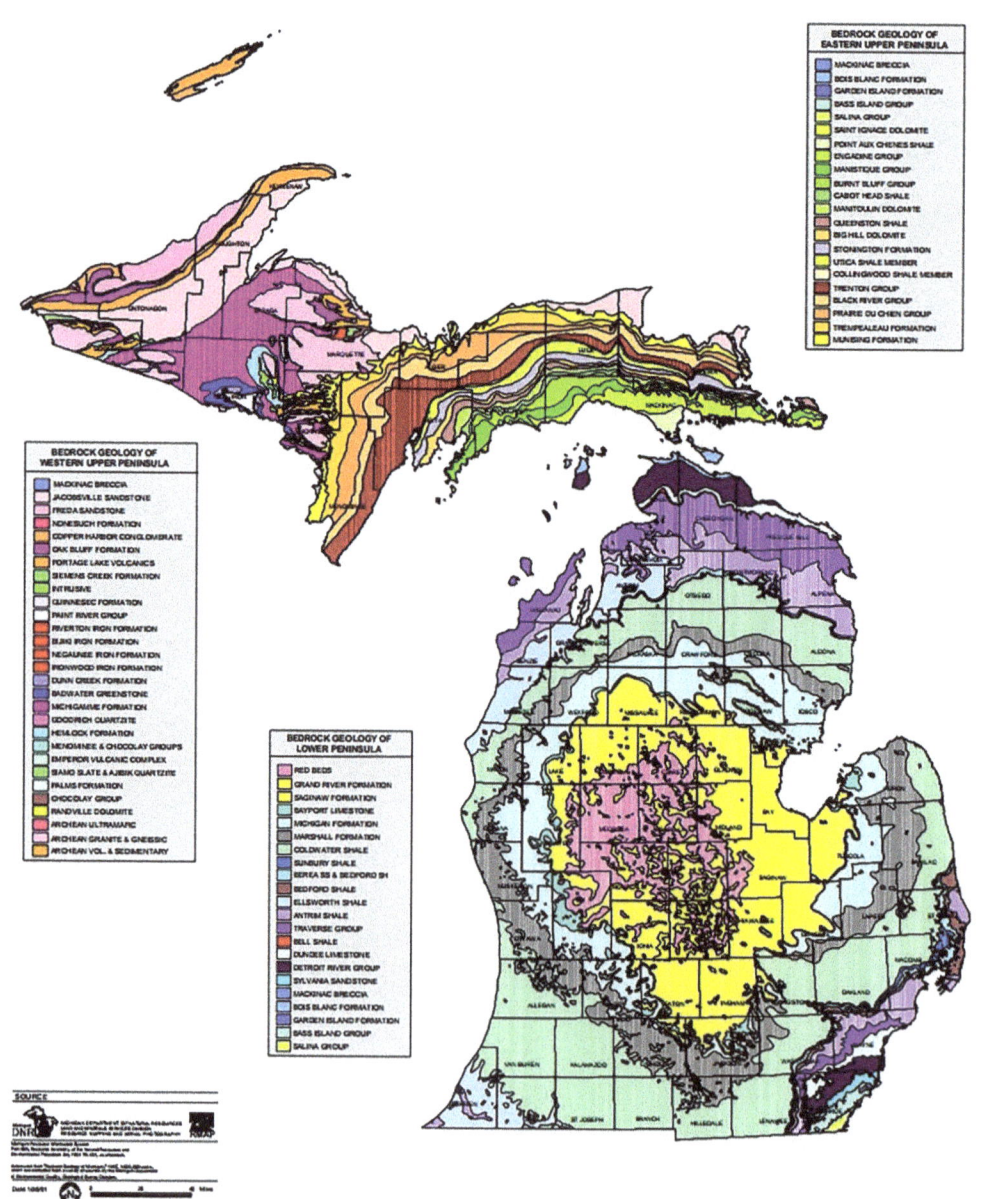

Fossil Cliffs at Grand Ledge

Ocqueoc Falls

Crossbedding at Grand Ledge

Petoskey Stone at Ocqueoc Falls

Notice the crossbedding found in the sandstone deposits at Grand Ledge. As I mentioned before, these are strong evidence that they were laid down in successive tidal events. Some polystrate fossils have been found here, though not of the size found at Joggins. A coal seam is found at the bottom of the cliffs. The cliffs at Oak Park in Grand Ledge are used by rock climbers for practice.

At the tip of the mitt from Charlevoix, Petoskey, Mackinaw City Rogers City and Alpena, limestone and coral fossils called Petoskey Stones and Charlevoix Stones can be found. Marine fossils are found throughout all of the deposits.

Ocqueoc Falls, near Millersburg west of Rogers City, flows almost entirely over Petoskey Stone. It is the only major waterfall in Michigan's Lower Peninsula.

In the Upper Peninsula, the Munising formation is classified as Cambrian. These are the cliffs forming the Pictured Rocks National Lakeshore. On top are glacial deposits. Every other layer in between is missing.

Crossbedding at Grand Ledge

Miner's Castle Before 2006

Miner's Castle in 2012

The bedrock geology of Michigan resembles a sliced onion, and the Munising formation extends all the way to Wisconsin where it forms the Wisconsin Dells. In the minds of a creationist, as these sediments were being laid down during the flood, the crust of the earth warped downward near the center of Michigan and formed a basin. Much like what we saw at Joggins, all the layers were shaved flat to the Jurassic by a runoff event, then glaciers carved out the basins for the Great Lakes, gouging through the softest of sediments. Glacial formations cover most of Michigan with the bedrock exposed only in a few places.

The Pictured Rocks is an example where catastrophic erosion has been observed. On April 13, 2006, one of the turrets of Miner's Castle collapsed and other rockfalls occurred in 2019 and 2021. These were witnessed by fishermen and tourists on the boat tours.

The boat tours originate in Munising and in the spring of the year, seasonal waterfalls cascade over the cliffs. Munising is the home of many waterfalls. They include Alger Falls, Wagner Falls, Horseshoe Falls, Munising Falls, Miner's Falls, Spray Falls and to the west, Laughing Whitefish Falls.

Pictured Rocks

Pictured Rocks

Pictured Rocks

Pictured Rocks

Pictured Rocks

Battleship Rock

Spray Falls

Spray Falls is just beyond Grand Portal Point. It is of interest because there is no eroded canyon leading up to its plunge over the precipice. Just beyond are the sand dunes at Grand Marias. Michigan is noted for its sand dunes and Lake Michigan has some of the best beaches in the world. The entire west coast of Michigan features sand dunes hundreds of feet tall. They range from the Indiana Dunes National Lakeshore in the south, Warren Dunes, Saugatuck Dunes, Hoffmaster State Park, Muskegon State Park, Silver Lake State Park, Mears State Park at Pentwater, Nordhouse Dunes at Ludington State Park, Orchard Beach State Park near Manistee, Sleeping Bear Dunes National Lakeshore, Fisherman's Island State Park and along US-2 in the Upper Peninsula just across the Mackinac Bridge.

On Lake Huron, Sleeper State Park and Port Crescent State Park features dunes. On Lake Superior, F. J. McLain State Park and Grand Marais have significant dunes. Dunes are fragile geologic ecosystems. An excellent explanation of the habitat of the dunes is given in a presentation at the Gillette Sand Dune Visitor Center at P. J. Hoffmaster State Park.

Surface Geological Map of Michigan

Much of the surface geology of Michigan is shaped by glaciers. It is here that creationist theory shines. During the flood of Noah, volcanic activity warmed the oceans as the continents split apart. Some who believe in an old earth object, saying that the energy to do so would have been enough to vaporize all of the oceans and kill all life on the planet. I am not sure what kind of scientific experiments they conducted caused them to make that conclusion. But the stratosphere and beyond into outer space provides a virtually infinite heat sink. Eventually the hydrological cycle kicks in producing vast amounts of rain and in the interior of the continents, snow, and ice. This would have kept the oceans from runaway heat. The edges of the oceans, even in the Arctic, would have been relatively temperate, but places like Michigan and Canada were affected by glacial ice.

Creation scientists like Michael Oard and Larry Vardiman believe that there was only one ice age which was part of the climate adjustment after the flood. The surface map of Michigan shows where lobes of glacial ice plowed through the fresh unconsolidated sediments laid down by the flood. Flood runoff would have scoured the tops of the bedrock sediment, and the glaciers would have ground out the basins for the Great Lakes. The northern Lower Peninsula exhibits evidence for this where Traverse Bay and Torch Lake are long scour marks. Many of the hills surrounding Gaylord are kames, eskers, and moraines, glacial features. Saginaw Bay was carved out by a giant glacial lobe and the Irish Hills in southern Michigan are glacial features.

Creationists argue that features like kettle lakes is evidence that the glaciers were recent and part of post-flood geologic events. Kettle lakes are formed from a block of glacial ice buried in sediment. At the Gerald E. Eddy Discovery Center at Waterloo Recreation Area, there is a trail that takes you to a bog where vegetation is growing over the top of a kettle lake. The fact that the kettle lake is gradually filling in at an observable rate indicates its youth.

The northern Lower Peninsula is abundant with kettle lakes. One of them, McCormick Lake, between Lewiston and Atlanta, is the source for the Thunder Bay River and just a little ways offshore it drops off to 50 feet deep. That area between Lewiston, Atlanta, Onaway, Rogers City and Alpena has a lot

Bog at Gerald E. Eddy Discovery Trail

of sinkholes with floating mats of vegetation gradually covering them and filling them in.

No discussion of Michigan geology would be complete without mentioning the Copper Country in the Keweenaw Peninsula. This is part of the Canadian Shield, composed of igneous and metamorphic rocks and is the source of copper and iron. For many years, copper and iron mines dominated

McCormick Lake, Montmorency County, a Glacial Kettle Lake

the economy of the Upper Peninsula. These rocks are supposed to be Precambrian, billions of years old according to evolutionary dating. Creationists would count these as rocks dating from the creation of the world. The Porcupine Mountains and surrounding areas form the highest elevation in Michigan. Mt. Arvon at 1,979 feet is the highest point in Michigan.

The Canadian Shield extends from the Northwest Territories, Nunavut, and Greenland in the north down into Michigan and northern Minnesota.

CHAPTER 32

Hawaii Volcanos

Here I am going to briefly mention a frequent argument that is given about the origin of the Hawaiian Islands.

According to evolutionists, the oldest of the islands is Niihau, Kalua, Lehua and Kauai, and the islands get younger as you proceed to the east. They date the eruptions Kauai to about 800,000 years old, but the big island of Hawaii is the youngest, with current volcanic activity, assuming uniformitarian erosion rates.

Hawaiian Island chain from Niihau to the Big Island of Hawaii

However, if you assume Flood Geology and Catastrophic Plate Tectonics, this plate would have moved rapidly during the flood, creating seamounts first from the Aleutians and like a zipper, erupting one by one in rapid succession. The first ones would be small, reflecting the rapid process. The islands started to erupt more continuously and get bigger as the plate movement slowed down.

Currently the hotspots in the island chain exist on the big island, and the new Kamaʻehuakanaloa (Loihi) seamount just to the east of the Big Island. So

Extended island chain and seamounts

the difference between the evolutionary explanation and the creationist explanation rests upon whether the plate moved at a slow rate or a rapid rate. Both explanations sound reasonable and in both cases the conclusions are dependent on the assumptions. We have no way to test this or draw one conclusion over the other given the forensic evidence we find in the present. Below is a map showing the entire seamount chain extending from the Aleutian Islands to Midway.

In the RATE study mentioned in an earlier chapter, samples of lava from several eruptions where the date of the eruption is known were taken to a laboratory where radioisotope dating was performed. Using the potassium-argon method, they yielded dates in the millions of years, which contradicted the known dates.

Extrapolating present rates of erosion into the past isn't a good indicator of the age of the islands. The island of Kahoʻolawe is in the rain shadow of Mount Haleakela on Maui and is a virtual desert, yet it is a relatively flat island. The rainy sides of Kauai, Oahu, and Maui, have steep cliffs, and this is an indication that those mountains are young. Mount Waialae on Kauai gets an average of 40 feet of rain a year.

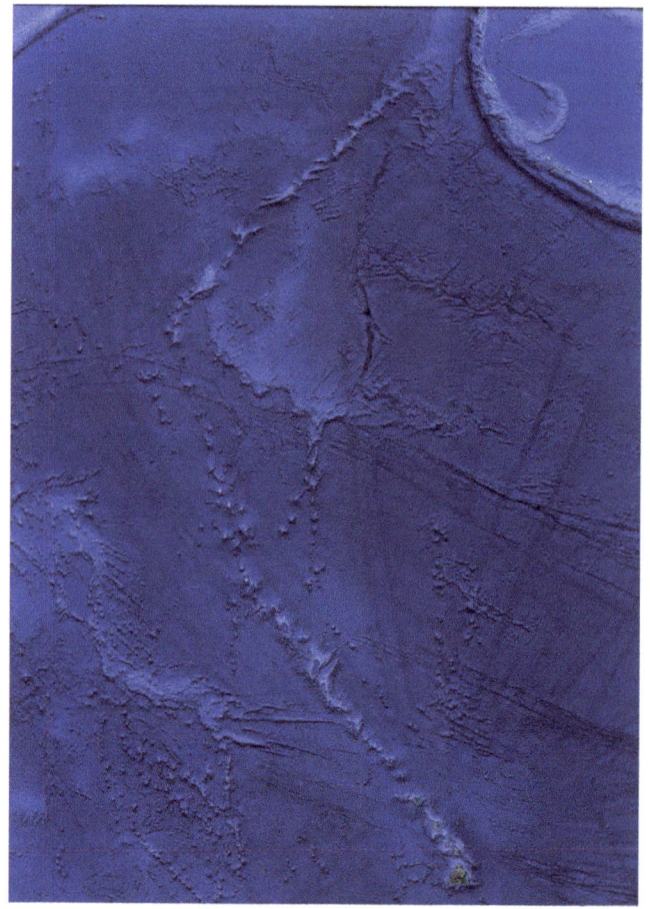

Extending the seamounts to the Aleutians

CHAPTER 33

Dinosaur and Human Tracks

Whenever you consider the idea that dinosaurs coexisted with humans, both before the flood of Noah and afterward, this creates a dissonance in the minds of evolutionary scientists.

Dinosaurs supposedly went extinct 65 million years ago according to the evolutionary geologic timeline. The Chicxulub impact crater found at the Yucatan is blamed for their extinction. So whenever dinosaur tracks and human tracks are found together, there is great effort on their part to discredit them. Dinosaurs are the poster child of the evolutionary paradigm and anything that suggests that humans lived at the same time would destroy their theory.

In 2009, we interviewed Dr. Carl Baugh at the Creation Evidence Museum in Glen Rose, Texas. He showed us his private collection of dinosaur and human footprints found at the Paluxy River. There is also Dinosaur Valley State Park nearby where, if you know where to look, you can find some of these. Of course, the State Park discounts the existence of human footprints and mocks the idea in their visitor's center. Certainly footprints are subject to interpretation, and we can really never use them much beyond Class B evidence.

The story of the Paluxy River tracks begins in 1908 when a massive flood ravaged the area, cresting at 27 feet. This flood removed layers of overburden exposing the tracks in the river bed. During times of drought, these tracks become visible, but as time passes, erosion has taken their toll on the tracks. A section of these tracks were excavated and placed in the American Museum of Natural History. Several of us participated in a dig in 2004 where we removed about a foot of overburden to expose an acrocanthosaurus track sequence where one

of them excavated earlier showed a human track inside. This track was called the "Beverly Track" after Carl Baugh's daughter and is on display in the museum. During our visit, the team of paleontologists uncovered ten new tracks in the sequence. This took place on private property near the banks of the Paluxy River.

Beverly Track

One rock slab that was uncovered showed four different tracks on top of each other from various species of dinosaurs. One of the dilemmas that faced us was whether to extract the tracks or leave them in situ. That track was buried again after being photographed and documented.

If a track is excavated, you lose context of where it was found, but if left in place, you risk exposure to the elements where it erodes. There is also risk of vandalism, and some of the tracks have shown signs of that. In all, 43 dinosaur tracks were exposed in this particular trackway.

Quad Track

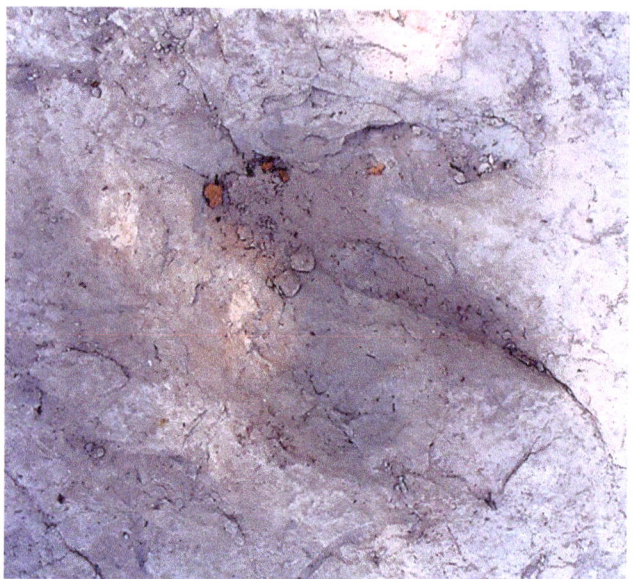

Acrocanthosaurus Track

Trackway Showing Where Overburden was Removed

This dinosaur track dig in 2004 was accompanied by several well-known creation scientists, and it was carefully documented so that nobody could accuse anyone of fabricating the tracks.

Digging up Dinosaur Tracks at the Paluxy River

The Delk Track

It was covered in mud and as he was cleaning off the dinosaur track, he exposed the human track. Dr. Baugh had the track examined with a spiral CAT scan, and the spot where the dinosaur track pushed into the human track showed compression, indicating that it was genuine.

Willett Track

In the private collection, the Delk track is one of the most spectacular. It is an acrocanthosaurus track stepping onto a human track. After a spring flood, Alvis Delk was collecting rock specimens near where White Bluff Creek empties into the Paluxy River. The flood had stacked up a pile of limestone blocks. Mr. Delk was a well-known fossil collector and kept this particular dinosaur track specimen in his home. After sustaining an injury, he needed some money for medical bills and decided to sell the track to Dr. Baugh at the Creation Evidence museum.

Another significant human track was discovered by O. W. Willett while he was fishing on the banks of the Paluxy River in the mid 1950's. He cut this fossil out of the ledge and took it home. This was before Dinosaur Valley State Park was established. This fossil features two tracks. The first track is that of a dimorphodon, showing left of center, with the hallux heel at the top. A dimorphodon is a flying reptile, a bit different from a pterodactyl. The human stepped upon the track afterward leaving a footprint showing distinct heel and toe prints. Leading up to where Mr. Willett cut this track out are a definite series of human footprints, and in 2009 we located them.

Stepping Stones to Park Ledge Trail

Footprint #1

Location of Park Ledge Trail

Compared to my foot

Human Track with toe prints

Stepping into the Track

Human Track Park Ledge Trail

Another Track

Dinosaur Tracks

A few steps away from the human tracks you can find dinosaur tracks on the ledge. An argument you can make for the tracks being genuine is the preponderance of poor tracks in the trails. If they were faked or carved, you wouldn't go to the trouble of making ones that looked bad.

During the depression in the 1930's many of these dinosaur tracks and a few human tracks were cut out of the rock and sold as souvenirs. Because of the effort required to cut them out, a few were carved. Because his brother Earnest "Bull" Adams had found a trail of dinosaur tracks and also human tracks, George Adams, after cutting them out and selling them, decided that he would attempt to carve a few. He gave up after making a few of them because the effort to do so wasn't paying off. This carving sat outside in the elements at his house and after his passing a crew was tearing down his house. During the demolition, Dennis Moore found this carving and brought it to the Creation Evidence Museum.

The Adams-Moore Track

This clearly shows the difference between a genuine track and a carved track. The chisel pock marks give it away, and there is no push-up on the sides of the track showing that the surrounding mud was displaced.

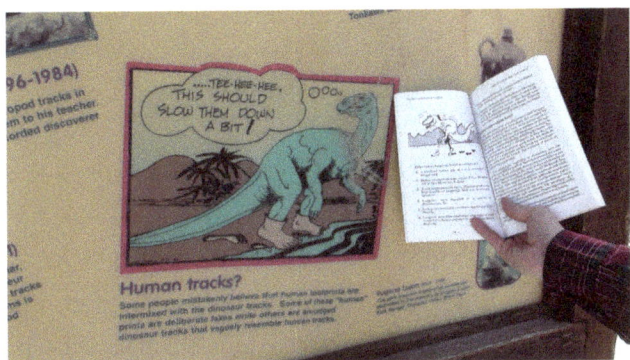

Sign Mocking Human Tracks

The State Park dismisses the human tracks as carvings or misshapen dinosaur tracks, mocking them in this depiction. Interestingly enough, they drew practically the same cartoon I use in my book *Revolution Against Evolution* to mock the position that the tracks are fake. The abundance of human trails with some crossing dinosaur trails, and in a few cases humans stepping into dinosaur tracks argue for them being genuine.

PowerPoint Slide by Don Patton

The locations of the Taylor trail, the Turnage-Patton trail, the Daugherty trail, the Beierle trail and others are clearly documented and can be viewed when the river is low. Dinosaur and human tracks have been excavated by removing over a foot of overburden.

William and Mabel Meister was searching for trilobites near Antelope Springs, Utah on June 1, 1968, when they found this moccasin print in shale. They cracked it open, and it revealed a trilobite that was crushed underfoot. The moccasin print shows stitching around the edge. This track is part of the collection at the Creation Evidence Museum.

Meister Print

Human Handprint in Carboniforous Rock

Compared with Hand

A human handprint was found in 1995 in Cretaceous rock, the same layer as the Paluxy tracks. It shows the webbing between the thumb and index finger, and an impression of the fingernail is found inside the hole where the middle finger dug its way into the mud.

A fossilized human finger was also found in the 1970's in Cretaceous rock. This finger was analyzed by spiral CAT scan, and it showed the bones and tendons inside.

Fossil Human Finger found in Cretaceous Limestone

There is a preponderance of evidence from Glen Rose, Texas that shows that dinosaurs and humans existed side by side. Since dinosaurs are supposed to be extinct 65 million years ago according to the evolutionary time scale, this evidence causes it to be called into question.

CHAPTER 34

Crater of Diamonds, Arkansas

The Crater of Diamonds is a state park near Murfreesboro, Arkansas. It is the only place in the world where the general public can participate in searching for diamonds. We talked to one fellow who works here ten hours a day, seven days a week sluicing diamonds and manages to make a reasonable living from his finds.

Plowed Diamond Field

Sluicing Diamonds

Strawn-Wagner Diamond Found Here

The land here is continuously plowed, and for an admission fee, you can explore the area and dig for diamonds yourself. Commercial mining was tried for several years at the beginning of the 20th century, but ultimately it was abandoned because the yield wasn't enough to make it viable.

Diamonds are found in kimberlite, pipes from volcanoes. Evolutionary geologists date these to billions of years old, however that has been called into question by creationists who, in the RATE study, subjected diamonds to Carbon-14 dating. Carbon-14 was found ten times that of the detectable limit, indicating that they are young. Some evolutionists will say that radiocarbon dating proves billions of years, but if they say that, they are either not understanding the method, or they are deliberately trying to deceive. The half-life of carbon-14 is 5,730 years, and the sample reaches below the detection level quickly if it is too old.

The early miners found coalified wood in the lamproite.[7] How could trees find their way in the middle of the volcanic layers? This is a mystery that evolutionists haven't solved.

CHAPTER 35

Smoky Mountains, Tennessee

Unless you actually study the geologic maps you find at the bookstore at the Great Smoky Mountains National Park, there is not much information that tells you that these mountains are upside down. But if you research the Great Smoky Fault, you will read a story geologists give you that there was a period of mountain building that took place millions of years ago when Pangea split apart, and the Precambrian strata was thrust on top of the Ordovician strata underneath.

The Precambrian is at the bottom of the traditional geologic timescale, and the Ordovician is supposed to be the age of the fishes.

The fault is at least nine miles wide and 248 miles long. It includes the Chilhowee Mountains to the north of the park, the length of the Foothills Parkway, and all the way to the Blue Ridge Mountains in the north. But in the middle are several valleys where the out of order sequence is exposed. They are Tuckaleechee Cove, Wear Cove, Cades Cove, and in a sinkhole near Crib Gap.[8] Geologists call these "fensters" or "windows", and they are the opposite of a klippe where the out of order contact is exposed around a mountain.

If you drive around the park taking Foothills Parkway to the south end, the fault is also exposed at Calderwood Gap. That drive has road cuts that expose vertical strata, bent strata, and tilted strata which indicate movement while the strata is soft and just laid down.

The difference between the evolutionary explanation and the creationist explanation is the time scale. Both of these explanations require that rock movement takes place. The creationist explanation that it took place over a short time helps explain fault gouge, slickensides, and striated rocks at the contact line, but also explains why the contact is sometimes found without those features.

The evolutionary explanation has the Precambrian rock strata creeping inch by inch on top of the Ordovician rock over millions of years. If Catastrophic Plate Tectonics is used as a model, you have sufficient momentum and force to make the strata reversals.

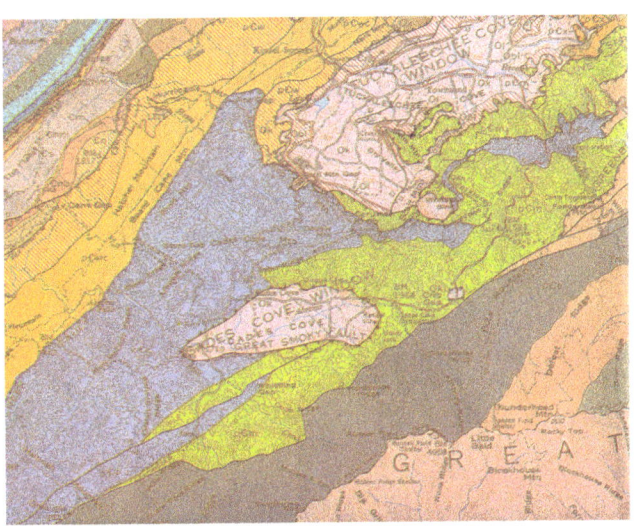

Cades Cove and Tuckaleechee Cove Window

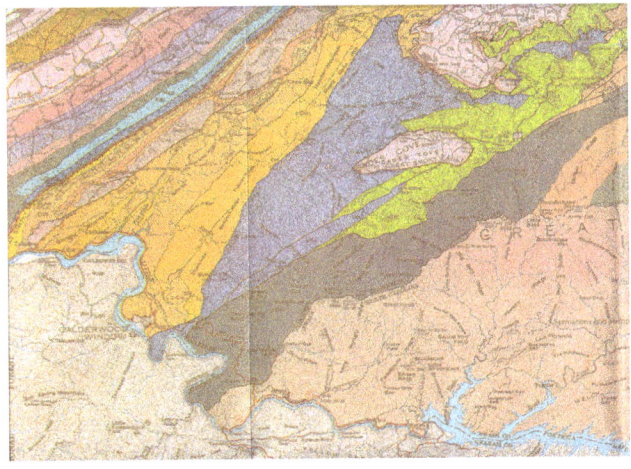

Chilhowee Mountain, Cades Cove and Tuckaleechee Cove

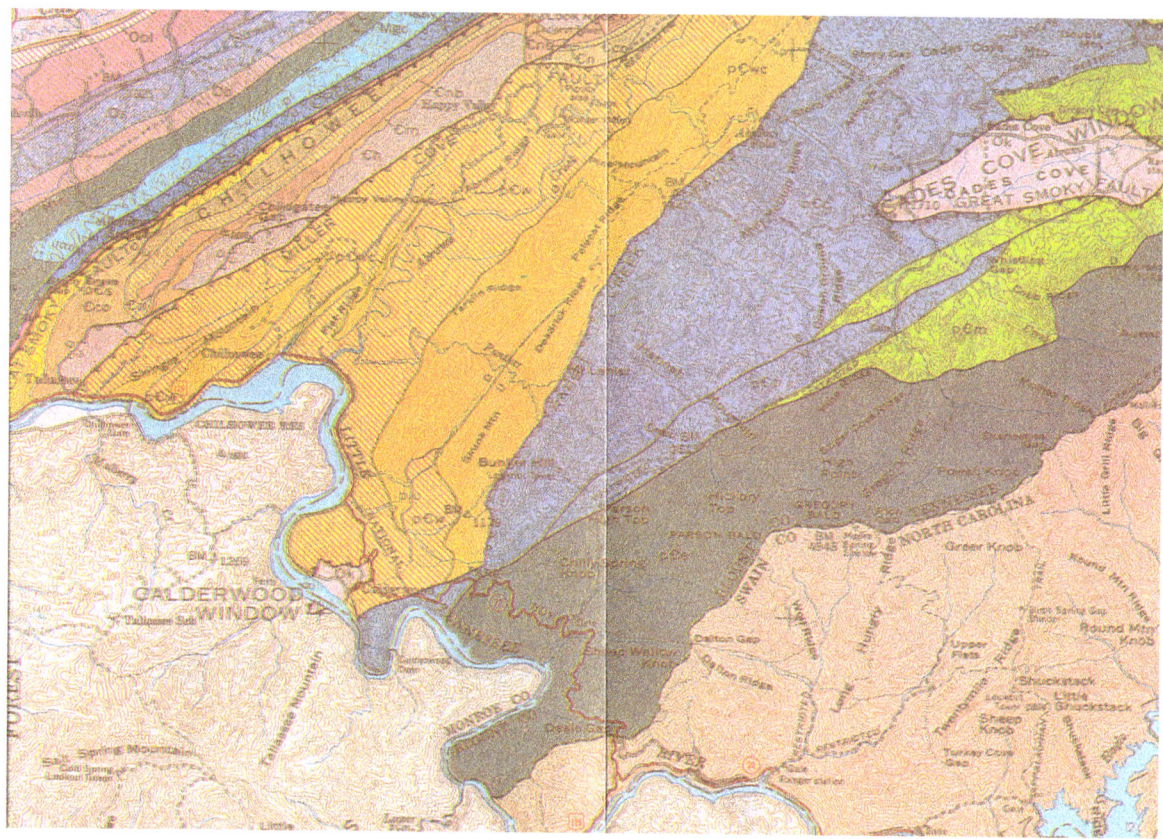

Calderwood Window

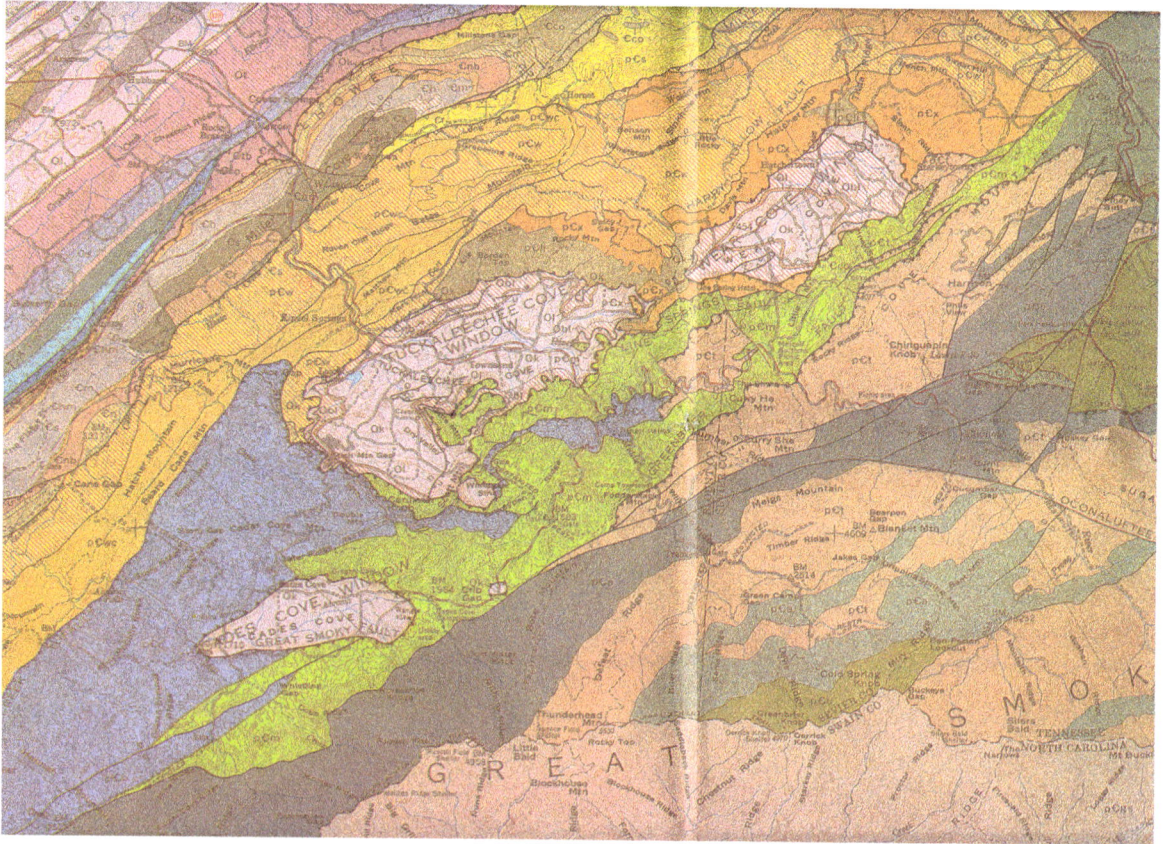

Cades Cove, Tuckaleechee Cove and Wear Cove

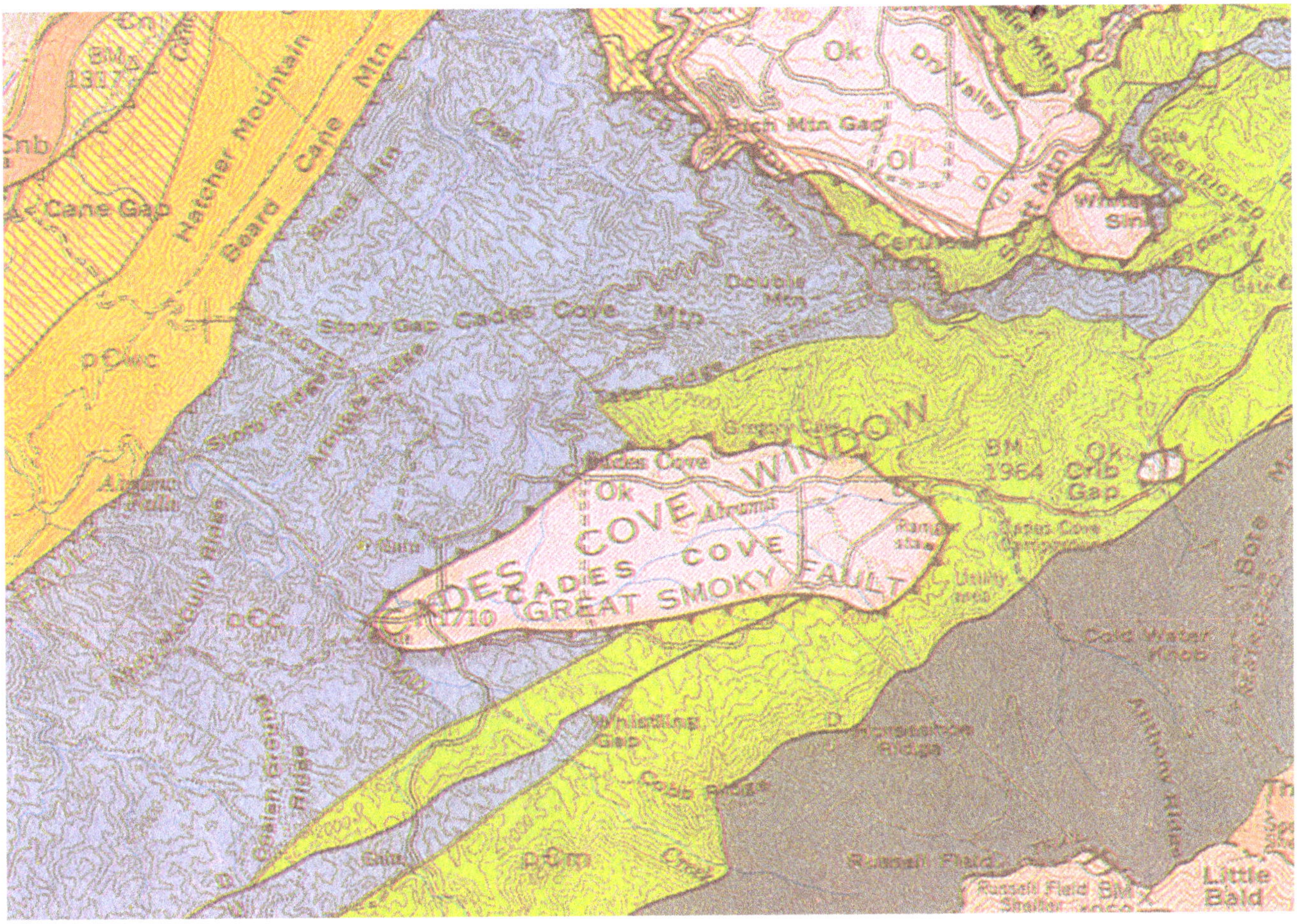

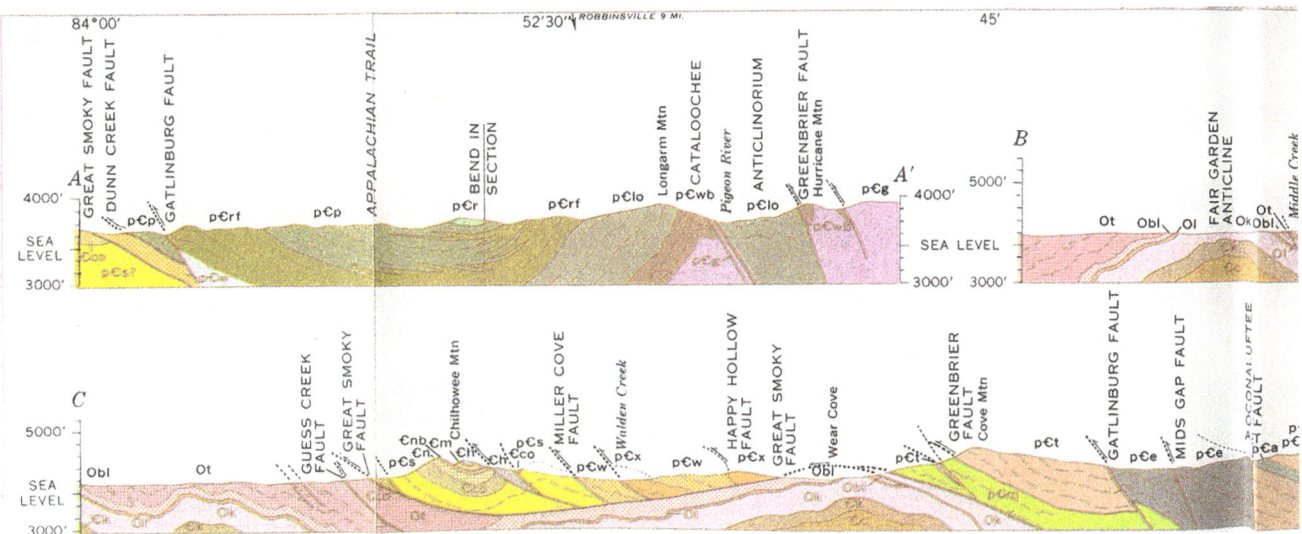

Cades Cove

Cross Section Geology of Smoky Mountains

Unlike the American Southwest, the geology of the Great Smoky Mountains is much more difficult to explore because the mountains are heavily wooded. Here we are displaying the geologic maps put out by the U.S. Geologic Survey and we are using their data and showing their interpretation. If you look at the cross section, the exposure of the Ordovician strata in pink with surrounding older rock gives you an idea how much it is out of order according to the traditional geologic column.

CHAPTER 36

Other Places to Explore

There are many more places of geological interest to creationists. Here are a list of what might be on my bucket list to visit or return to see again because my photographs are so old they have deteriorated.

Yosemite National Park, California

This park is an example of post-flood ice age geology where U-shaped valleys are carved out by massive ice sheets. The High Sierras provided a barrier to the weather engines that carried evaporation from the oceans warmed from the aftermath of volcanic activity during the flood. This piled up snow and ice creating rivers of glaciers that carved out Yosemite valley, sheared Half Dome and left hanging valleys such as where Bridal Veil Falls hurtles over the cliff.

Lassen Volcanic National Park, California

Lassen, in northern California is an example of an active volcano in the Cascades range that last erupted in 1915. Vestiges of that eruption are still evident at the park with fumaroles and hot spots still active. Though the eruption occurred in 1915, potassium-argon dating of the Mt. Lassen plagioclase yielded dates in the range of 0.11±0.03 million years ago. This was part of the RATE study.

Crater Lake National Park, Oregon

Crater Lake is a caldera that formed after the volcano erupted. On the north end of the lake, there is a trail that takes you to a boat dock where you can cruise on the lake and observe dike intrusions in the cliffs such as one called the "devil's backbone" and Phantom Ship. The boat will take you to Wizard Island where you can climb to the top of the cinder cone. The boat may also take you up close to the "Old Man of the Lake", a floating log that bobs vertically, weighed down by rocks embedded in its roots. This essentially is an example of the floating log mat theory proposed by creationists as an explanation of the formation of coal. Also of note is that the mountains in the area, Union Peak and Mt. Thielsen, are steep-sided volcanic plugs where the softer sediments have been eroded away. At the south end of the park, the Pinnacles are columns hardened by fumaroles where the surrounding dirt has been washed away.

Mt. St. Helens, Washington

The 1980 eruption of Mt. St. Helens provided a great example of the effects of catastrophic events that occurred during Noah's flood. The bottom of Spirit Lake is an example of coal formation with the floating log mat above. Samples from the lava dome yield radioisotope ages of about 350,000 years, far too old. Creationists have written much about Mt. St. Helens and there is a museum called Mt. St. Helens Creation Center dedicated to the creation explanation in Castle Rock, Washington.

Yellowstone National Park, Wyoming

This is the ultimate young earth volcanic park with Old Faithful, the Yellowstone River Gorge, Yellowstone Lake, and all of the geysers. The so-called multiple successive forests of Specimen Ridge trips up some people, but if you view it in the context of a massive flood depositing trees vertically in layers all in one event, this makes much more sense. It really poses the same problem that Joggins presents to evolutionists: the fossil trees wouldn't have survived years of forest succession to leave fossils. A fossil only occurs when it is buried quickly.

Fossil Butte National Monument, Wyoming

This place is the source of finely graded layers of sediment called "varves" which evolutionists like to count up as millions of years. But because fish are buried in these varves, and they are buried polystrate in a bunch of them, it is more reasonable that all of these were laid down in one event. You can

purchase a toy that has sand in water, and if you tip it over, the sand cascades through the water to form finely graded layers. You can see a display of these at the Ark Encounter that explains them.

Nebraska Sand Hills

There is a vast area between the Niobrara River to the north and the Platte River to the south in Nebraska that is made up of sand hills that appear like ripple marks in the landscape. This is clear evidence of flood runoff.

Dry Falls, The Palouse, and the Missoula Flood, Washington

Eastern Washington and western Montana were studied by J. Harlen Bretz in the 1920's and he concluded that a breached ice dam inundated the entire area, causing the scablands called the Palouse and poured over a cliff called "Dry Falls". Bretz was ridiculed for his ideas, but later when he was 96 years old, he was awarded the Penrose medal for his research. During the ice age, which, according to creationist models, was just after the flood, a giant lake formed behind the ice sheets, and when it was breached, it carved out these channels.

The Hell Creek Fossil Beds, Glendive Montana

Site of major dinosaur fossil finds, including ones where soft tissue is found.

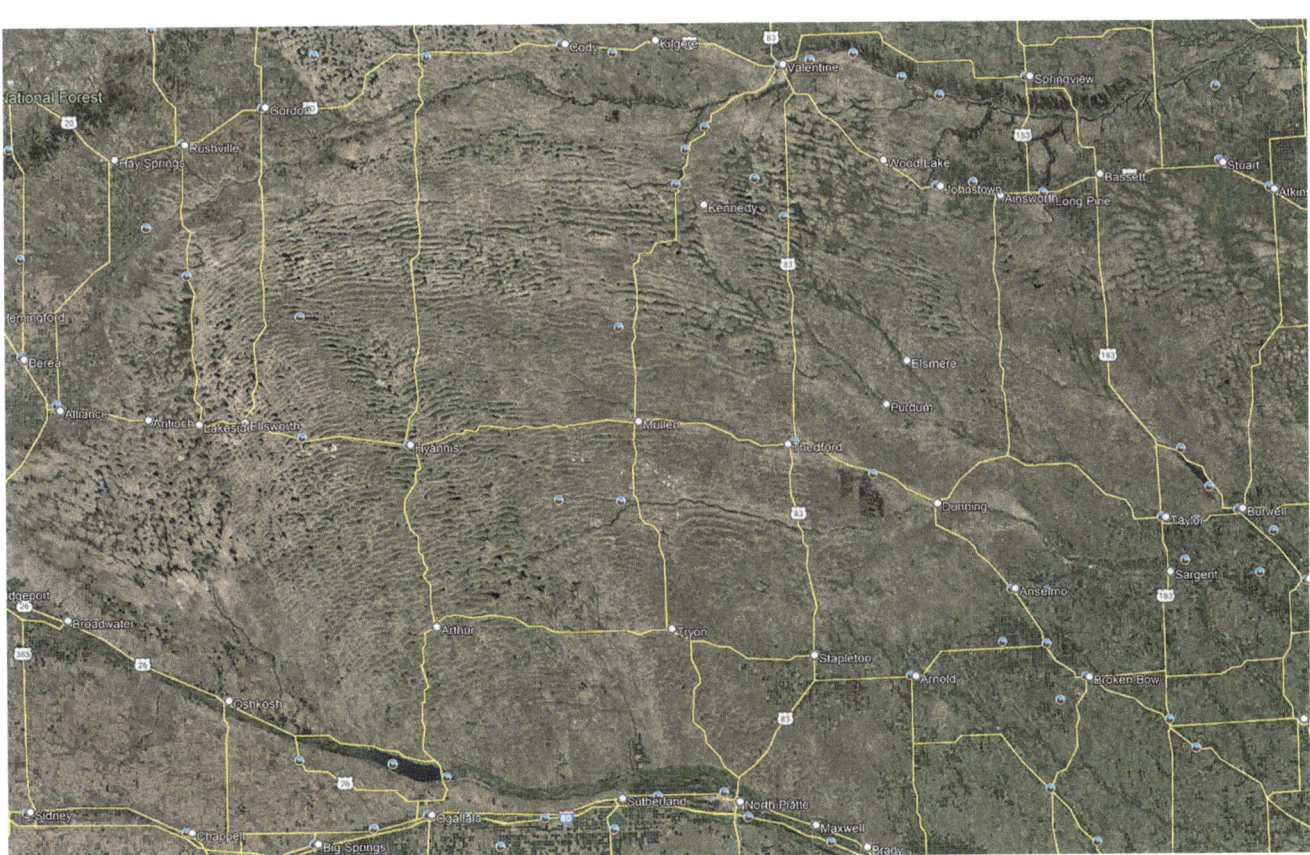

The Nebraska Sand Hills

Drumheller Fossil Beds, Alberta

Home of the Big Valley Creation Science Center. Many dinosaur fossils found here are found as bone, not fossilized.

CHAPTER 37

Conclusion

Creationists and evolutionists formulate their models of origins based upon the same physical evidence found in the rocks. What is vastly different are the starting assumptions. In both cases, the underlying worldview determines the conclusions. If we put on biblical glasses through which we view the world, we get our clues about the geology of the world through two distinctive events: the original creation in the beginning, and the resurfacing of the earth that took place during Noah's flood. The Bible lays out a timeline in history when this took place through the genealogies of the patriarchs.

Evolutionists, because of their predisposition to reject both of these historical records in the Bible, have bought into the idea that the rock layers were laid down over millions of years. Creationists believe that the worldwide sequence of sedimentary rock points to one event that wiped out the face of the earth and through successive tidal events buried the plants, animals, and vegetation deep below layers of sediment during catastrophic plate tectonic continental movement. Reasons for this belief are numerous.

- Vast reservoirs of fossil gas and oil are 12,000 feet deep in the earth's crust.

- Out of place fossils.

- Fossils that extend through several strata (polystrate trees).

- Overthrusts, out of sequence strata exposed with klippes and windows.

- Paraconformities.

- Parallel strata with smooth contact lines.

- Seismites.

- Piping.

- Crossbedding.

- Carbon-14 is found in diamonds, coal, and limestone.

- Sedimentary rocks on the continents but mostly volcanic on ocean basins.

- Evidence for catastrophic erosion of the Grand Canyon and surrounding areas such as Monument Valley.

- Superior explanation for the formation of coal with floating log mats, as demonstrated at Spirit Lake near Mount St. Helens.

- Coal seams buried thousands of feet.

- Vast areas of strata sliced off.

- Strata on continents, not ocean basins.

- Meager river deltas.

- Mountains with concave strata.

- Michigan's onion bedrock geology

Because we cannot go back into the past and retrieve the conditions under which fossils, strata, mountains, and oceans were formed, we are left with what we observe in the present. We can only guess how it all came to be. Empirical science is observable, testable, and repeatable. It is not applicable in the case of geology since the experiment cannot be conducted or repeated. The method of science used is forensic science, and it has its limitations.

Creationists have developed a good working model of how a global flood as described in the Bible could have produced the geology we see today.

Other Books by Douglas B. Sharp

Endnotes

1   Steno, N. The Earliest Geological Treatise—1667, translated by Axel Garboe, MacMillan & Co. London, p. 17, 1958

2   https://www.icr.org/RATE

3   https://answersingenesis.org/store/product/grand-canyon-different-view/

4   https://answersresearchjournal.org/grand-canyon-monument-ancient-earth/

5   https://answersingenesis.org/geology/mount-st-helens/ argon-in-mineral-concentrates-from-mount-st-helens-volcano/

6   https://cronkitenews.azpbs.org/2018/12/21/phoenix-valley-geology/

7   Worthington, Glenn W. Genuine Diamonds in Arkansas, Mid-America Prospecting, 81 Roy Road, Murfreesboro AR., p. 128-129, 2003

8   https://www.nps.gov/parkhistory/online_books/geology/publications/pp/587/sec2.htm

Geology Theology Photo and Illustration Credits

Geology Theology Photo and Illustration Credits
LR – Linda Rusiecki
RG – Rich Geer
GF – Guy Forsythe
DV – David Vonderheide
EGLE – mi.gov/egle
USGS – US Geological Survey
All others taken by Douglas Sharp

GT02 – LR
Page 18 Pu'u O Kila Lookout, Na Pali Coast, Hawaii
GT81a – Google Earth
Page 45 Zion National Park Aerial View
GT81GrandStaircase – Taken at Geology Center
Page 45 The Grand Staircase
GT94 - GT100 – GF
    Page 53 Seismite
    Page 54 Parabolic Recumbent Fold
    Page 54 Lizard Head Recumbent Fold
    Page 54 Water Escape Slits
    Page 55 Another Seismite
GT103WupatkiDragon – DV
    Page 68 Puff, the Magic Dragon
GT402Glacier – LR
    Page 160 Divide Mountain (in the distance)
GT402GlacierChiefDiagram – RG
    Page 161 Chief Mountain Overthrust
GT404 – GT408Glacier (21) – LR
    Page 162 Running Eagle Falls (Trick Falls) in
        2021 at peak
    Page 162 Local Wildflowers
    Page 162 Hiking to a Glacial Lake
    Page 162 Driving to Logan Pass
    Page 163 MacDonald Falls
    Page 163 Avalanche Lake
    Page 163 Trick Falls, Trick Falls
    Page 163 Arete
    Page 164 Upper MacDonald Creek
    Page 164 Glacier Flow
    Page 164 St. Mary's Lake and Wild Goose
        Island
    Page 165 Beargrass
    Page 166 Wildflowers
    Page 166 Mountain Vista
    Page 167 Waterfall
    Page 167 Waterfall
    Page 167 Avalanche Falls
    Page 167 St. Mary's Lake
    Page 168 Lake MacDonald
All other photos taken by Doug Sharp
GT415 – EGLE
    Page 178 Michigan Bedrock Geology
GT426 – EGLE
    Page 181 Surface Geological Map of Michigan